W9-ABL-562
7/07
Fayetteville Free Library
300 Orchard Street
Fayetteville, NY 13066

100 Butterflies and Moths

PORTRAITS FROM THE TROPICAL FORESTS OF COSTA RICA

THE BELKNAP PRESS OF HARVARD UNIVERSITY PRESS

Cambridge, Massachusetts ▪ London, England ▪ 2007

100 Butterflies and Moths

JEFFREY C. MILLER
DANIEL H. JANZEN
WINIFRED HALLWACHS

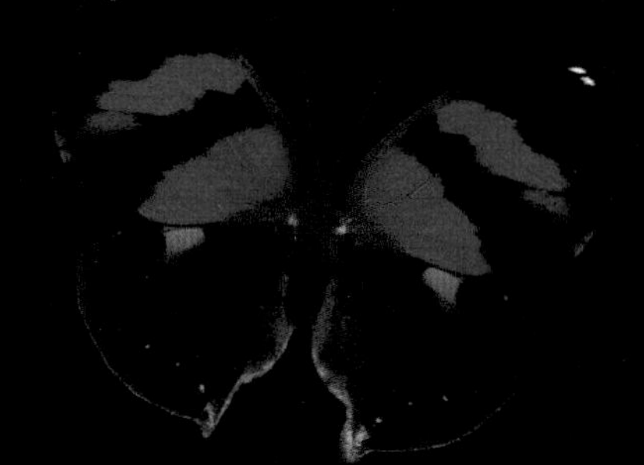

Image on pp. ii–iii: *Belemnia trotschi, Siderone marthesia* (inset)

Printed in the United States of America

Library of Congress Cataloging-in-Publication Data

Miller, J. C. (Jeff C.)
100 butterflies and moths : portraits from the tropical forests of Costa Rica / Jeffrey C. Miller,
Daniel H. Janzen, Winifred Hallwachs.
p. cm.
Includes bibliographical references and index.
ISBN-13 978-0-674-02334-5 (cloth : alk. paper)
ISBN-10 0-674-02334-X (cloth : alk. paper)
1. Butterflies—Costa Rica. 2. Moths—Costa Rica. 3. Butterflies—Costa Rica—Pictorial works.
4. Moths—Costa Rica—Pictorial works. I. Title : One hundred butterflies and moths.
II. Janzen, Daniel H. III. Hallwachs, Winifred. IV. Title.

QL554.C8M55 2007
595.78′9097286—dc22 2006051111

This book is dedicated to the expert taxonomists of the world.

You are the plant and animal scientists who make possible books like these.

Contents

A Natural History of Butterflies and Moths 1

A Gallery of Butterflies and Moths 19

Behavior, Ecology, and Caterpillar Images 121

Life History and the ACG Environment 221

What's in a Name? 231

The Parataxonomists: Lepidoptera and Plant Biologists 235

References 243

Acknowledgments 249

Numerical Species List 253

Alphabetical Species List 255

A Natural History of Butterflies and Moths

A typical response to a museum tray of moths and butterflies, the colors and patterns resplendent against a black velvet background, is to classify them on a subjective scale of beauty. Perhaps such a conclusion is a result of the presentation, or perhaps it is because moths and butterflies neither sting us, alarm us, nor give us indigestion (given that we do not eat them). We generally do not become alarmed at the sight of a butterfly or moth since there is no fear of impending discomfort. Therefore, we have not developed an impression of "good" versus "bad." In fact, those species that advertise themselves with brilliant colors, colors that signal to other animals that gastric distress will be the order of the day if eaten, we consider beautiful, desirable. Thus we regard each species as an object of beauty—its colors and patterns invoking our aesthetic sense. It is notable that in North America more species of butterflies are on the Federal List of Threatened and Endangered Species than any other insect. Beauty attracts attention. The judging of beauty is a matter of personal interpretation, but the design behind the perceived beauty is a product of evolution, a dynamic, ongoing process involving many variables.

Other species are also attracted to and react to moths and butterflies. Predators, parasites, and mutualists all use color and pattern as a means of distinguishing friend from foe from food. Each of these species finds its own meanings and exhibits different reactions to the colors and patterns of moths and butterflies.

In this book, we present 100 moths and butterflies from the Area de Conservación Guanacaste (ACG), a conserved wildland in northwestern Costa Rica with an area equal in size to that of New York City and its suburbs. The species were chosen from among 10,000 species of moths and butterflies estimated to exist within the ACG to represent taxonomic diversity, variation in behavior, life-history tactics, and also what has been reared to date in the nearly three decades of caterpillar inventory conducted in the ACG (for an account of the caterpillar inventory, see our previous book, *100 Caterpillars*, and a later section in this book, "The Parataxonomists"). We chose to show these moths and butterflies through one of many possible lenses, as artistic portraits pinned against a black background, partnered with species accounts that illuminate their life histories, trusting that a beautiful creature is made more beautiful if it is understood. Our presentation focuses on color and pattern, but there are so many other ways to know a butterfly or moth—through flight, size differences, nuances in wing positions, wingbeats, smell, wind, sun, air temperature, and sound.

The images cannot substitute for a field experience, but not everyone can achieve the field experience. The species accounts that follow the portraits in this book provide the armchair equivalent of such an experience. As biologists, we try to grasp the natural history and evolutionary processes that produced the details of the butterflies and moths we see. A foraging bat sees a moth as a speeding, slowing, darting, angling, spiraling blur on its mental sonar screen. A jacamar, a Neotropical bird that specializes in catching diurnal butterflies, sees it as a colored blur of potential food or as an object of frustration and futility because its prey is just plain too fast. A flower regards butterflies and moths as pollen couriers that can be bribed, or perhaps enticed, to assist in the act of reproduction through an offering of nectar, or conversely as nectar thieves. Very likely many species do not interact with butterflies and moths at all.

MOTH AND BUTTERFLY WINGS

Moths remain motionless during the day, whereas butterflies remain motionless during the night. The ramifications of this overly simplified dichotomy between closely related insects, both of which develop from a caterpillar, are in part manifested in the colors and patterns of the wings. It is these colors and patterns that catch attention or result in the individual being overlooked, and therein lies the basis for the evolution of two basic survival tactics, crypsis and aposematism.

Did you ever wonder why the wings of butterflies and moths are proportionately so much larger than those of other insects? The large wing size is not just for lifting the body into flight, for that can be accomplished with smaller wings and a faster wingbeat. Lepidoptera transmit a variety of messages through their wings based on color and pattern. Both the upper (dorsal) and under- (ventral) sides of their wings may be encrypted. The messages that are conveyed have evolved to include hiding, warning, and attraction. Big wings provide a larger billboard for sending messages to potential mates and possible enemies.

The dorsal side of a butterfly's wing is often dramatically different in color from the ventral side, both because each wing surface may be displayed for different purposes and because the combination of the two in flight produces yet a third color pattern. Compare the dorsal views of the butterfly species shown here with their ventral views shown in the individual portraits listed in the caption. Why are the various species designed so differently? We will provide some answers to that question.

The wings of some Lepidoptera have evolved to be seen as part of the background. This strategy, called crypsis, permits the species to blend in and hide while in plain sight. Crypsis is certainly a risky survival tactic to employ in the face of potential

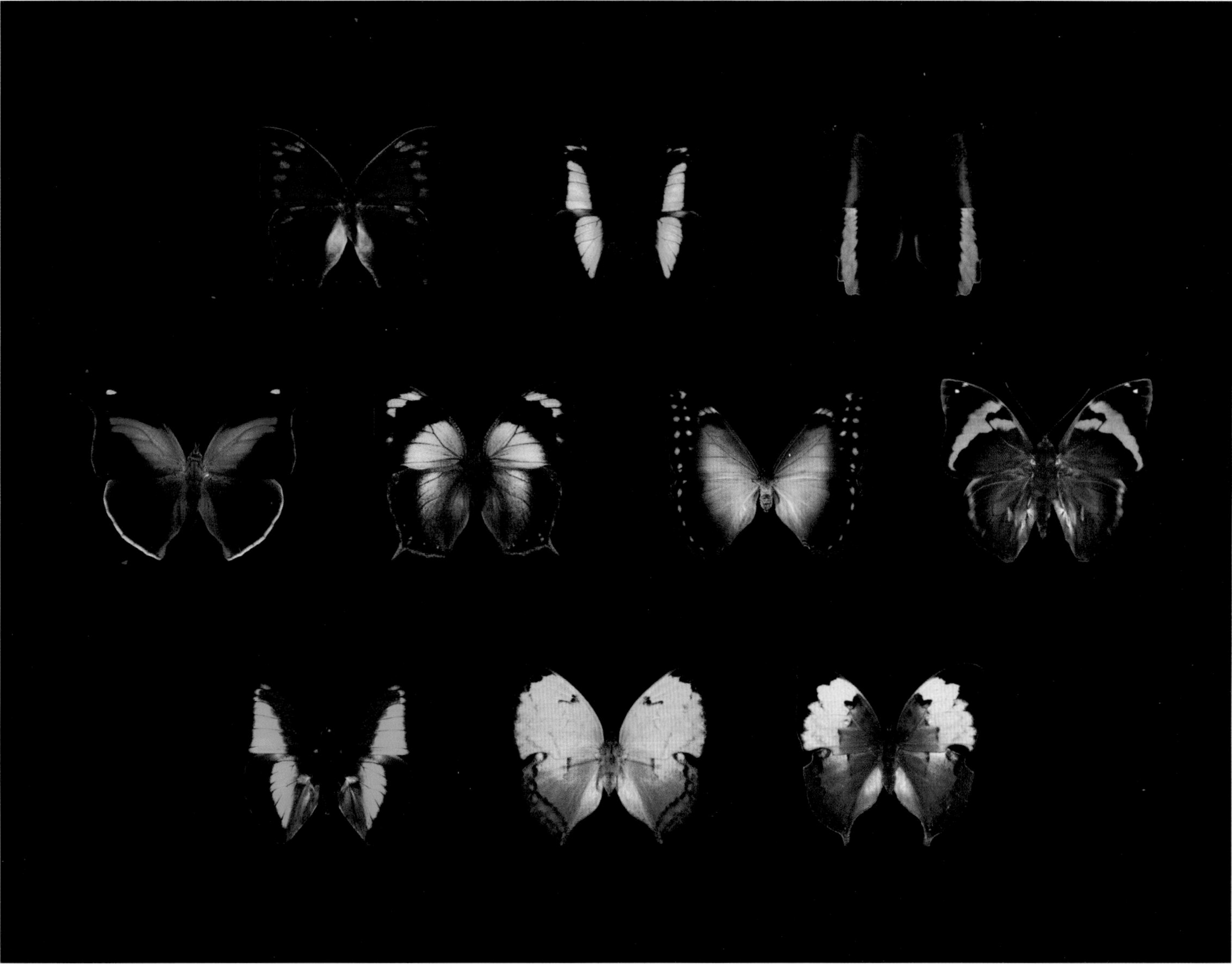

Dorsal surface of wings: ventral surfaces are shown in the Gallery of Butterflies and Moths following this chapter. Shown left to right, top row: *Anaea aidea,* #26; *Archaeoprepona demophon,* #79; *Caligo atreus,* #36. Middle row: *Historis odius,* #25; *Memphis proserpina,* #40; *Morpho theseus,* #33; *Opsiphanes cassina,* #35. Bottom row: *Prepona laertes,* #80; *Zaretis ellops,* #69; *Zaretis itys,* #70.

predators. What if the trick should fail? If a species is to evolve into adopting the tactic of crypsis for survival, then it had better be good at hiding.

In contrast, the wings of other species of moths and butterflies have evolved to be seen from afar. The colors and patterns beg for attention, denoting the presence and exact location of the individual. This sort of flag-waving is a defensive technique founded in part on aposematism, the survival tactic of being ostentatious and obvious. Aposematic species typically exhibit some combination of two or more colors that likely include iridescent blues or greens; brilliant reds, oranges, and yellows; and white. Often the background to the bright colors is black, or there is a noticeable black accent. Black serves as a stage for the brighter colors and enhances their contrast, making very clear the signal regarding the ecological function that the colors depict, such as, "I am poisonous, do not eat me." The attention-grabbing colors may be signals to mates, but most often we interpret them as warning signals to would-be enemies, either relaying a message about its bad taste or toxicity, or mimicking a species that relays such a warning. Our eyes and brain are drawn to bright colors and striking patterns probably because of their coincidental resemblances to the colors that have really mattered to us during our evolutionary history—fruit, prey, blood, and courtship.

The analogy that the wing of a moth or butterfly is like a painter's canvas is relevant when kept in the context of design. Designs on the wings become even more remarkable when viewed in the context of natural selection and evolution. The most obvious design element of butterfly and moth wings is the bilateral symmetry between the left and right side. That is, the forewing on the left is a mirror image of the forewing on the right. The same holds for the hindwings. Experienced lepidopterists will recognize that this rule is broken in specimens that are gynandromorphs, male on one side and female on the other, therefore not bilaterally symmetrical. Gynandromorphism is particularly notable in species expressing sexual dimorphism. Gynandromorphs aside, bilateral symmetry is the rule in the overall design of wings. Note that in some species, the forewing and hindwing are more or less similar in pattern and color, but in others the forewing and hindwing are drastically different. This is demonstrated in many species of *Automeris* (#3, #4, #85, #86), in which the hindwing is normally tucked under the forewing when the moth is at rest. When an intruder ventures near, however, the hindwing with its brightly colored eyespot flashes into view.

A comparison of dorsal and ventral views also reveals another element of design. In some species the dorsal and ventral wing surfaces are similar, but in other species the two surfaces are drastically different. In fact, the divergence in pattern on the dorsal and ventral surfaces is most prevalent among the butterflies. In such species it is the ventral surface that displays a more intricate combination of colors and patterns. This may seem surprising at first, but it makes sense when paired with the realization that butterflies are primarily diurnal species. They typically perch with their wings held together over their bodies, exposing the entire ventral surface of the hindwing and a small portion, the distal tip, of the ventral surface of the forewing. In this position it is the ventral surface, mostly of the hindwing, that is exposed for other organisms to see.

Again, the adaptive function of wing colors and patterns becomes glaringly apparent when you compare the exposed ventral tip of the forewing in certain species of butterflies and notice how it does not match the central and basal area of the ventral forewing colors and patterns (e.g., *Mesotaenia barnesi*, #99 and *Opsiphanes cassina*, #35). Instead, the ventral surface of the tip of the forewing matches the colors and patterns of the ventral hindwing. As noted, the patterns match because the larger forewing extends beyond the smaller, but overlying, hindwing when the butterfly is perched and is therefore exposed. Not all species show extreme divergence in color across the entire surface of the forewing venter (e.g., *Memphis proserpina*, #40 and *Historis odius*, #25), suggesting that some aspect of these species' perching behavior is different. All, of course, are examples of natural selection at work.

Each adult has a dorsal side that it presents to one audience, a ventral side that it presents to another, and a combination of the two in flight that it presents to yet another audience. Each design displays the adaptive measures best suited to that species.

MOTHS

What is a moth? It is an adult metamorphosed from a caterpillar, no longer just a feeding unit but a mating and dispersal unit. As such, it is adapted to avoid predators that hunt by sight while it passes the period from dawn to dusk. Some moths are active during the day, and not all moths keep busy all night. There are distinct times of courtship, oviposition, feeding, and migrating. For example, there are at least fifteen species of *Automeris* (Saturniidae, the giant silk moth family) in the ACG, and at least another six overlooked species "hiding" within them. It is likely that not only does the female of each species produce a different pheromone to attract its respective males, but also the virgin females of each species probably call (release pheromone) at a different time of night. Despite this variation, it appears that each species of female saturniid lays her eggs during the same period of only several hours shortly after dusk.

We, as diurnal and large visually-orienting predators, see moths perched motionlessly in the wild much in the same way as they are found, or more likely missed, by foraging birds and monkeys. Night-foraging birds, bats, frogs, scorpions, and spiders likely have an altogether different impression. The great majority of moth species appear to us as selection has driven them to appear—inert and inedible objects, tree bark, lichen, twigs, dead leaves, green leaves, stones, dirt, and fungus. It is amazing how many ways natural selection can make a live animal look like a dead leaf or twig.

A few moth species have gone the other way, displaying ostentatious colors and behaviors that signal either a toxic lunch or a stinging insect, or its harmless mimic. These moths usually fly by day. Although these species are often lumped with butterflies by the casual observer, which is not surprising since they have been created by a regime of diurnally foraging predators, they are not in the six families formally labeled butterflies: Nymphalidae, Papilionidae, Pieridae, Riodinidae, Lycaenidae, and Hesperiidae.

Each major lineage of moths has spun off brightly colored, diurnally active species, two of which we feature: *Erbessa salvini* (#65) and *Zunacetha annulata* (#66). Also, sometime in the distant past, at least two major lineages of moths became what we today call butterflies. Whether any of the living moths and butterflies resembled these deep ancestors we will leave to the phylogeneticists to determine. However, what we can say is that when a moth evolves into the diurnal world, the game of survival changes dramatically from one of hiding from predators all day to one of being a flying, flapping, sailing, perching billboard. The contents of this book are an ode to those signals, and human fascination with these colors and patterns is the reason why the pages highlight so many butterfly species, so far out of proportion to the numbers of butterfly species in the ACG.

BUTTERFLIES

The transition to a diurnal world is not an easy one. When a species evolves to follow a pattern of diurnal activity, it moves into: 1) a very dry world where flight pumps dry air through the body, taking away moisture that must be recaptured at some point (in contrast to flight at night, in what is often nearly 100 percent relative humidity); 2) a world where a flying object can be seen at tens of meters, within the foraging range of vastly more predators than the resting moth, 3) a world where large, flapping wings that beat only a few times a second, instead of buzzing moth wings, scream to both mates and potential assailants, and 4) a world where the wings must serve a double function—courtship and predator avoidance. With a few conspicuous exceptions, nocturnally active moths probably do not much incorporate their wing patterns and colors in their courtship displays, and do not use these colors and patterns to escape from nocturnal predators.

The challenge of being a tropical butterfly in a dry and hot diurnal world has been largely unstudied. In the ACG it is notable that a large number of species of rain-forest butterflies, presumably evolved for the relatively humid daytime rain forest, do not enter the ACG dry forest, even though there are species of food plants in the dry forest that appear adequate for development of their caterpillars. As Lepidoptera move into the diurnal world, they must begin the arduous and risky task of locating water to replenish that which they have lost through respiration. Searching for water leads to many other activities, such as flower visitation and puddling, in which males perch on mud and suck up moisture and minerals. Diurnal Lepidoptera also have improved eyesight, presumably to keep an eye on potential predators or court potential mates. Many moths have well-developed eyes as well, but their superior eyesight is used to see flowers at night and perhaps to aid oviposition.

When the butterfly is in flight, neither the potential predator nor other butterflies see it as we see it here, in the flat and two-dimensional view of a preserved specimen. As the wings flap, the butterfly alternates showing the dorsal, then the ventral surface of the wings. The visual impact is a complex combination of colors and patterns. The color of a butterfly in flight is further complicated by the angle of the sun, shadiness and sunflecks in the forest understory, and the background vegetation. It is extraordinarily difficult to imagine how the colors of any particular butterfly shown in the pages of this book contribute to the actual visual impression we receive in the butterfly's natural surroundings.

Two examples will suffice. The gray-white half moon on the forewings of *Porphyrogenes* BURNS01 (#46), as seen in our image and in the drawers of museum specimens, looks quite different in situ. If you were to hold the specimen so that you see the wing at a very shallow angle, simulating the dim light of dusk or dawn, the two half moons turn into a pair of brilliant white reflectors, not unlike those put on jogging shoes. A second example is our photographs of the dorsal side of the wings of *Siproeta epaphus* (#31) and *Adelpha basiloides* (#32), both diurnal Nymphalidae occurring in the same rain-forest habitat. As one of many members of the genus *Adelpha, A. basiloides* is involved in a huge mimicry complex. Butterfly workers know both of these butterfly species well, but rarely think of them as mimics, because when pinned and spread, the pattern of red-orange, white, and brown is not at all similar. *Siproeta epaphus* also has two to three times the wing area of *A. basiloides.* But watch them in flight at any distance and each is an organized blur of white, orange, and brown that can very easily be viewed as belonging to the same species. Who is the mimic and who is the model? It is unknown—perhaps both are mimics and models.

LIFE CYCLE AND BEHAVIOR

Though all travel from egg to larva (caterpillar) to pupa to adult, moths and butterflies exhibit a wide array of behaviors within each life stage, all naturally selected for survival and thus the perpetuation of the species. The life cycle of *Rothschildia lebeau* (Saturniidae), which occurs in the ACG dry forest, provides an excellent model of complete metamorphosis. A newly eclosed female perches on vegetation while copulating with a male (not visible in the illustration on page 8). Her wings, resembling nothing more than a dead, fungus-ridden leaf, protect her from the attention of predators. She will likely lay her eggs on the first night following mating in a clutch of eight on the underside of a leaf margin. The first instar (not shown) emerges from the egg, feeds for two or three days, and molts into a second instar. The second instar is brightly colored and spiny, mimicking an urticating limacodid or megalopygid caterpillar. The third instar (not shown) molts into the fourth instar, also mimicking an urticating limacodid caterpillar but much larger and differently colored than the second instar. The fifth and last instar is finely haired and now a twenty-gram caterpillar, mimicking an urticating hemileucine saturniid or toxic sphingid caterpillar of equal size. The silk cocoon is spun by the last instar during the prepupal phase. While weaving the cocoon, the caterpillar incorporates an escape hatch, located at the top end where the cocoon is anchored to the branch, for the emerging adult.

One evolutionary goal of survival is to get the opportunity to mate. On page 9, we show a pair of *Eacles imperialis* on the morning after the female emerged from her cocoon, still copulating many hours after sperm has passed from the male. The female is stuffed with several hundred eggs that are ready to be deposited, whereas the slender male, mimicking yellowing leaves, does its best to avoid a vertebrate predator. Certainly this is a very precarious position to be in for such a long time. He will remain with her until the late afternoon, maximizing her chances of survival. If a predator finds her, half the time it will grab him instead of her. Furthermore, if he were to leave her now, it would not increase his chances of finding another mate, so he has more to gain by protecting his genes.

Following a successful mating, the female adult deposits eggs—sometimes singly, sometimes in clusters or masses. Typically, moth and butterfly eggs are glued directly onto the foliage of the caterpillar food plant. Their colors and positioning are a combination of crypsis, aposematic warning colors, mother's flight mechanics, and serendipity. Although very few insect species guard over their eggs, not all eggs are unprotected. In the top left corner of the illustration on page 10, a silk-

Lepidoptera life stages as illustrated by *Rothschildia lebeau:* adult, eggs, second instar, fourth instar, fifth instar, pupa.

Mating *Eacles imperialis*

covered egg mass of *Hylesia lineata* (Saturniidae) is covered by a felt of partially barbed abdominal hairs. They severely irritate sensitive skin if contacted. Often eggs are deposited onto the undersides of leaves. In the upper center photo, the tip of the abdomen of *Caio championi* (Saturniidae) is reaching under the leaf to place her black-ringed white eggs in a loose and unorganized cluster. Many species will place their eggs on a very specific plant part, as in the case of *Pseudbarydia crespula* (Noctuidae) (above, right) who carefully puts her eggs at the tip of an expanding *Inga* leaf. A female *Eumorpha satellitia* (Sphingidae) (center, left) placed a single egg at the tip of an expanding *Cissus* leaf just after the rainy season had started. The entire egg load from one female *Arsenura armida* (Saturniidae) may total over 300 eggs placed on the underside of a mature leaf of *Pachira quinata* (center). The timing of oviposition must be just right in certain cases. In the center right photo, an *Aellopos titan* (Sphingidae) egg was placed on a thorn on *Randia* in anticipation of the bud burst that accompanies the first rains after the long dry season. The evenly spaced orange eggs of *Melete isandra* (Pieridae) (below, left) were craftily positioned on a mature leaf of *Phoradendron* (mistletoe is toxic) and probably are aposematic. Also deliberately tightly packed are the black-spotted white eggs of *Periphoba arcaei* (Saturniidae) on a mature leaf of *Pachira quinata* (below, center). The black spots on the top of the eggs may mimic parasitoid exit holes, thus deterring an actual parasitoid. Lastly, the orange eggs of *Lonomia electra* (Saturniidae) (below, right) were carefully spaced in distinct rows on a mature leaf of *Citharexylum*.

The larval stage of life, the caterpillar, is divided into instars demarcated by the event of molting. Our first book, *100 Caterpillars: Portraits from the Tropical Forests of Costa Rica*, focused on this life stage, so we defer for now to the pupal stage. Pupae come in many forms, wrapped in silk (a cocoon) or naked, cylindrical or bulging, immaculate or marked in a variety of ways, on vegetation or in the soil, and possessing colors of all sorts. The more spectacular pupae are those of butterflies. Butterfly pupae, sometimes called chrysalids or chrysalises, are exposed and usually hang head down and naked from a pad of silk into which the tiny hooks of the cremaster, the basal tip of the chrysalis, are entangled. The wing pads wrap around the ventral side

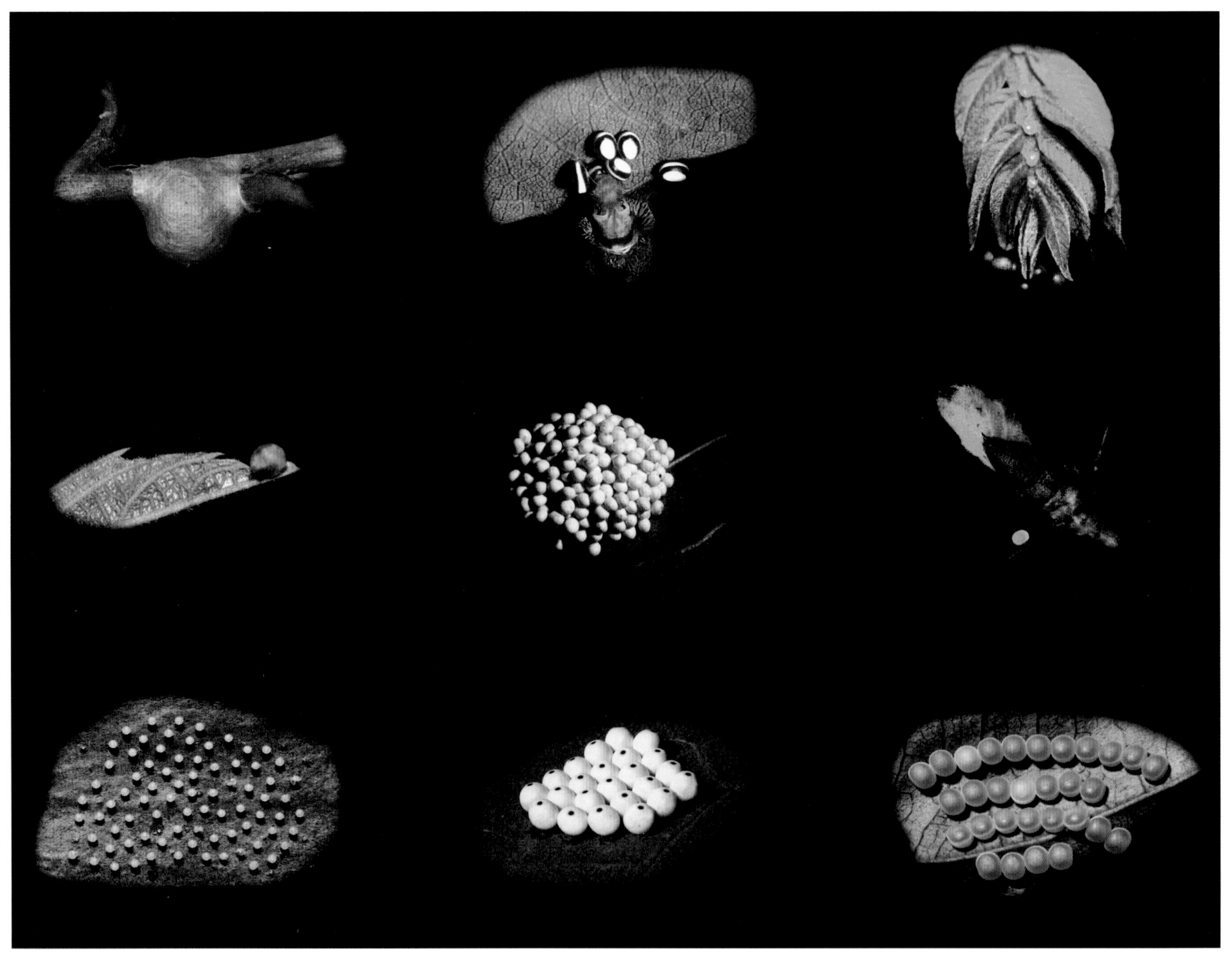

Eggs of Lepidoptera. Left to right: top row—*Hylesia lineata, Caio championi, Pseudbarydia crespula;* middle row—*Eumorpha satellitia, Arsenura armida, Aellopos titan;* bottom row—*Melete isandra, Periphoba arcaei, Lonomia electra.*

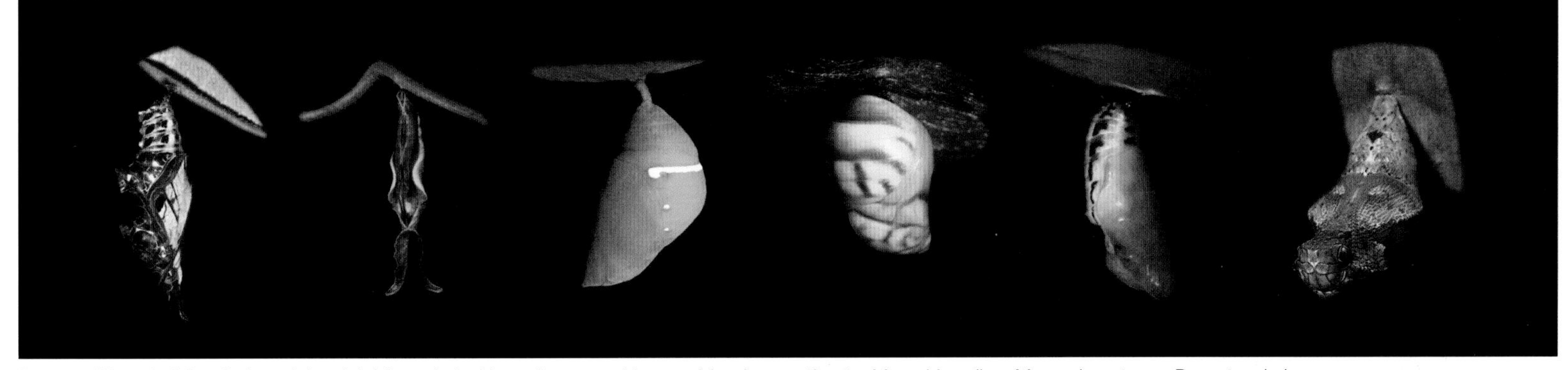
Pupae of Nymphalidae. Left to right: *Adelpha celerio, Hamadryas amphinome, Morpho amathonte, Memphis aulica, Marpesia petreus, Dynastor darius.*

and the abdominal segments are externally visible. This is in strong contrast to the black to dull brown of most moth pupae, often hidden in a silk or silked leaf cocoon in the litter on the forest floor or underground. Associated with the fact that butterfly pupae are attached to vegetation and exposed is the exhibition of crypsis and aposematism. Just as for caterpillars and adults, pupae exhibit warning colors and mimicry. Six examples of nymphalid chrysalids are shown above in our composite plate of pupae. On the far left is the shiny and silvery-reflective *Adelpha celerio*, simultaneously presenting an appearance of a torn leaf and a bright sunspot shining through a leaf. The floppy ear-like antennae of *Hamadryas amphinome* look like a dying shriveled leaf. The bulging, velvet green *Morpho amathonte* gives the impression of a slightly damaged leaf, whereas the bulging beige-tan *Memphis aulica* appears to be a tree snail mimic. The chrysalis of *Marpesia petreus*, shiny, seemingly moistened, and bright yellow with black markings, resembles a slug. Lastly, and most remarkably, is *Dynastor darius*, a snake-head mimic.

Despite the simplicity of the life cycles we have presented above, from egg to larva to pupa to adult, moths and butterflies display a wide variety of unusual behaviors. As shown in the photos on page 12, a male *Copiopteryx semiramis* (Saturniidae), an occupant of the ACG dry forest, is an excellent mimic of a dead leaf. Perched on foliage, it displays sixteen-centimeter-long tails that look like leaf petioles. At night, in flight, the tails make the moth seem larger than it really is on a bat's radar screen.

Next, a black-winged skipper, *Carystus phorcus*, displays the tell-tale red compound eyes characteristic of nocturnally or crepuscularly active species.

An undisturbed and undescribed female, very similar to *Gamelia musta* (Saturniidae), blends quite well on a bed of dead leaves, but after being disturbed, she maneuvers her forewings to flash a bright red spot on the hindwing—a behavior meant to startle the intruder.

On page 13, a *Copaxa moinieri* (Saturniidae) moth failed to startle off the intruder but did survive an attack. The female dis-

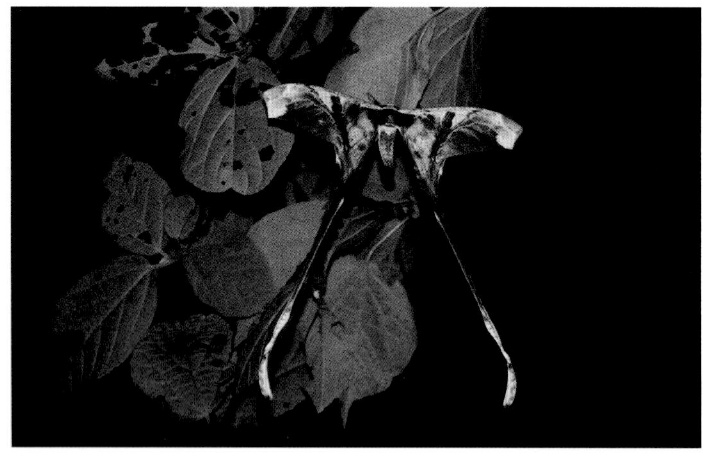

Copiopteryx semiramis

Carystus phorcus

plays the monkey bite marks that pierced her wings; a nearly fatal encounter, but she is still capable of flight.

Opposite, the female of *Orgyia* JANZEN01 (Lymantriidae) is wingless (apterous). She mates and lays her massive clump of eggs while remaining stationed at the exit hatch of her cocoon. In apterous species, dispersal is performed when first instars are blown by the wind or crawl in search of a suitable food plant.

Gamelia JANZEN01, resting undisturbed

The identical *Gamelia* JANZEN01, after being disturbed

Copaxa moinieri

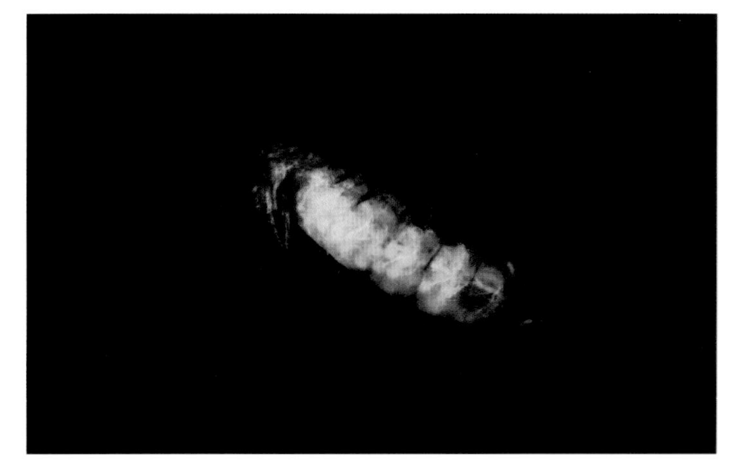

Orgyia JANZEN01

Another mode of transportation is to be phoretic, that is, hitch a ride on another individual. A female *Ptilopsaltis santarosae* (Tineidae) firmly attaches to the fur of a mouse, *Liomys salvini*, thereby getting a ride into the mouse's underground nest, where the moth will lay eggs. The resulting caterpillars, feeding on refuse, will clean up the nest.

A male *Protographium agesilaus* (Papilionidae) is puddling, another method of acquiring what might be considered uncon-

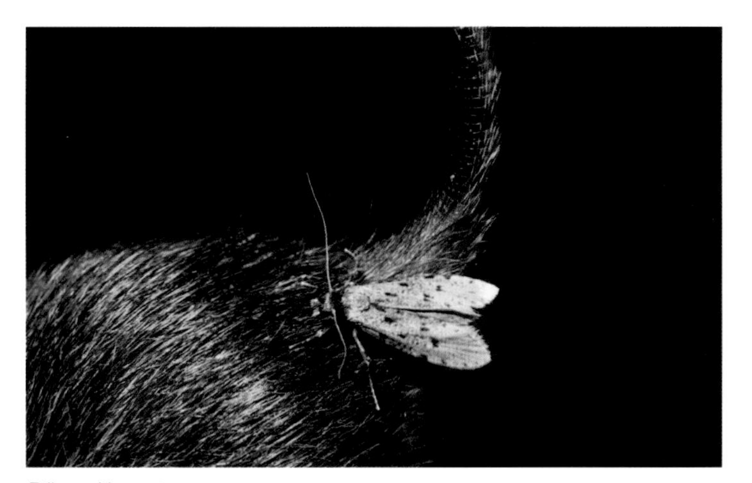

Ptilopsaltis santarosae

Protographium agesilaus

ventional food. He seems to be drinking, but in fact he is sucking up large volumes of water, which he quickly excretes to filter minerals and enhance his chances of passing on his genes.

Another use of a probing tongue is to suck up nectar and collect pollen. In the photo on this page, a rain-forest *Heliconius* sp. (Nymphalidae) encrusts its tongue with pollen from cucurbit flowers, pollen from which the amino acids are being leached and then in turn sucked up. Note that butterflies have a straw-like proboscis, or tongue, and cannot chew or masticate solid food.

In addition to a fascinating collection of behaviors, the variety in the expression of color and design is one of the many reasons why the study of Lepidoptera is enjoyable. Tracing the evolution of these colors and designs often lays bare the species' fascinating natural histories. Consider the accompanying figure illustrating three pairs of species, a male and female of each, showing the divergence of color and pattern based simply on

Heliconius sp.

gender. From left to right, we show *Morpho amathonte* (Nymphalidae), *Parides iphidamas* (Papilionidae), and *Phoebis philea* (Pieridae), with the female always appearing in the top row. Many male moths and butterflies have different color patterns and wing shapes than the females for obvious courtship reasons, but also because they often live largely in different microhabitats, doing different things. Thus, each sex is subjected to somewhat different pressures of natural selection, particularly based on predator regimes.

The real puzzle in Lepidoptera biology is what, when, where, why, and how these colors serve any given species in its daily tasks of survival and ultimate goal of reproduction. Although crypsis and aposematism are the tools, determining when, with whom, and how they are used is not an easy task.

Will the real nonmimetic and nonaposematic butterfly please raise its antennae? No takers, as we suspected. One advantage of working with a large collection such as the ACG inventory is the ability to see similarities and differences in all species at the same time, rather than focusing on one family of butterflies or moths, as is the usual practice of taxonomists and collectors. One such realization is that just about every species of butterfly and day-flying moth appears to be aposematic (warning-colored) or a mimic of either another butterfly or moth or an inanimate object such as a dead leaf (see *Zaretis*, #69, #70; *Memphis*, #39, #40; and *Anaea aidea*, #26). There are hundreds of species that look like other hundreds of species. Hesperiidae, of which we feature fifteen species, are particularly noticeable in this respect. The ACG inventory has now reared over 400 species of skipper butterflies. If we were to consider

Examples of sexual dimorphism. Females on top row, males on bottom row: left to right—*Morpho amathonte, Parides iphidamas, Phoebis philea.*

them in the way a predator might see them, these hundreds of species can be condensed to about ten somewhat overlapping color morphs: brown with long hind-wing tails, small and dark brown, small and orangish yellow brown, white, brown with blue undersides, brown with white or cream spotted undersides, brown with white tails, brown with a yellow posterior part to the hind wings, blue/black/white striped, black with a red-orange patch, and scattered among them, a few conspicuous mimics of butterflies in other families.

Today the ACG species, with their attendant morphologies and behaviors, express the results of the process of natural selection. This process started somewhere else in the tropics long before the terrestrial component of the ACG emerged from the ocean. Certain volcanic and tectonic events had to occur before these species could come together as we see them now.

ORIGINS OF THE AREA DE CONSERVACIÓN GUANACASTE

Costa Rica, a land bridge between South and North America, was once submerged. The position of Costa Rica puts it in a fascinating biological location in the context of animal and plant movement, both colonizations and extinctions, between two major continents. The region encompassed by the ACG is so biologically rich, yet it is much younger than the terrestrial regions to the north and the south. Certainly geological age is not the determining factor providing for such biological diversity, but the geological history of the region helps to explain where the species came from.

A million and a half years ago, Lake Nicaragua extended 100 kilometers south of its current southern shore to the present-day site of the Costa Rican town of Liberia in Guanacaste Province. On its eastern shore was a 3,000-meter tall volcano that erupted in one Krakatau-like event and filled the Costa Rican part of the lake with 200 meters of rock and ash. Then, merely 50,000 years ago, the Rincón de la Vieja complex of volcanoes emerged within the original crater. Today this complex of volcanoes rises to 2,000 meters. At the northern end of the complex, Volcán Cacao sprouted about 30,000 years ago. Then to the north of that, just 20,000 years ago, Volcán Orosí arose. Together these volcanoes form the Cordillera Guanacaste, a mountain range that forms a wall between them and the Pacific lands to the west.

The Area de Conservación Guanacaste. North is to the left. The red higher-elevation areas highlight the Cordillera Guanacaste.

The formation of the Cordillera Guanacaste changed the regional climate. The moisture-laden trade winds, blowing from the Caribbean, hit this wall, rise, condense in the cold upper elevations, and drop their moisture. Thus the Cordillera creates a rain shadow, resulting in the relative dryness of the Pacific side of the mountain range, the dry forest. Today's ACG slices across this diverse contour of Caribbean rain forest. The Cordillera itself is crowned by cloud forest and the dry forest of the rain shadow on the Pacific side. About as many species of moths and butterflies live within this transect, perhaps 10,000 of them, as occur in continental North America north of Mexico.

So, the short answer to where the moths and butterflies of the ACG came from is that they immigrated from all directions

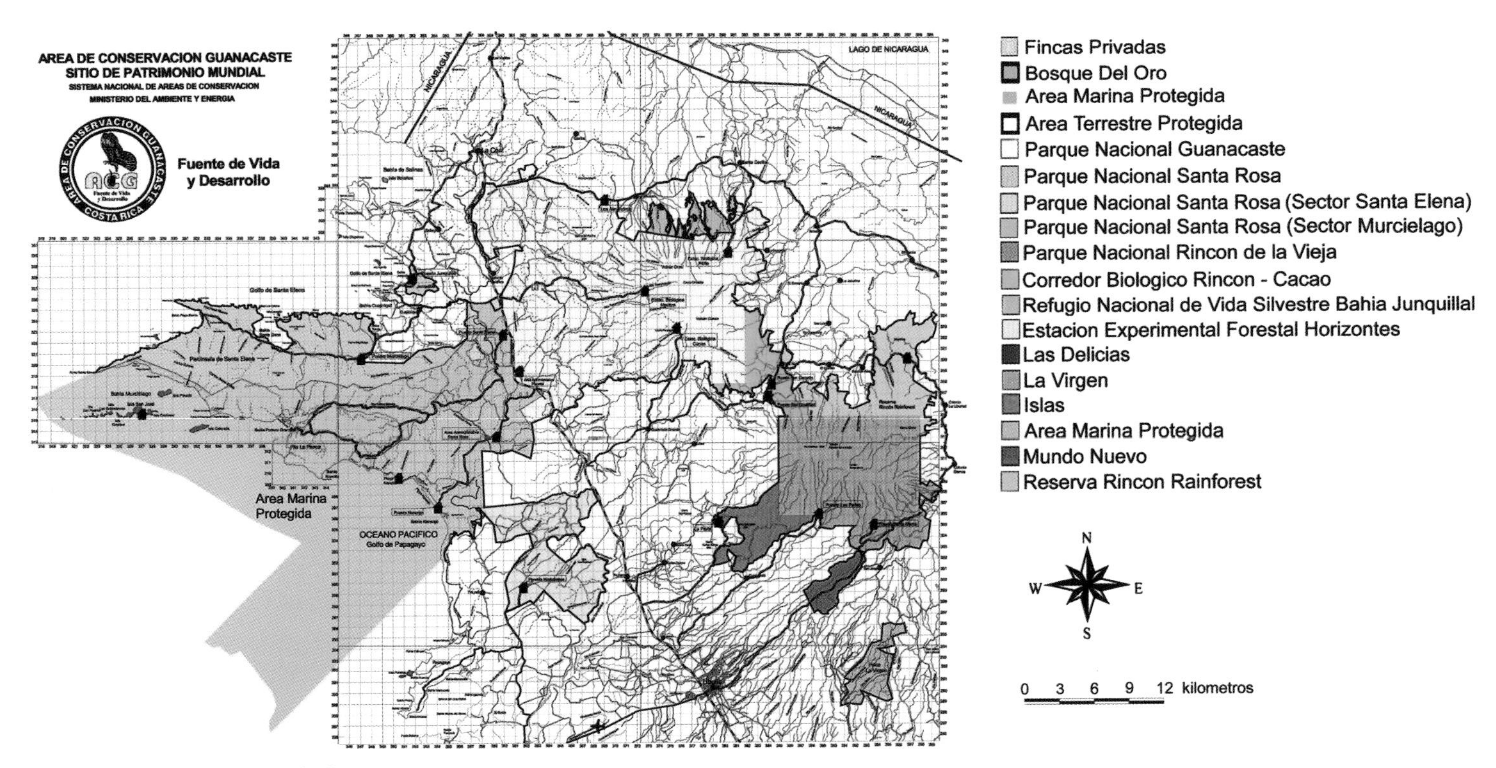

Map of the Area de Conservación Guanacaste.

1.5 million years ago, and then reinvaded on a smaller scale as the Cordillera Guanacaste took shape. The ACG was never an island unto itself but rather a "melting pot" of immigrant species, essentially all outsiders arriving in their own ways and ecologically fitting together with their traits, for better or for worse.

Since the primary "ecological islands" in the ACG—the upper elevations of the three volcanoes—are only 20,000 to 50,000 years old, it is unlikely that there has been enough time or distinctive eco-evolutionary pressure for all but a small number of new species to evolve on them and generate what are commonly called "endemic species." These volcanoes were probably colonized in great part by upper-elevation species arriving from other highland areas by their own muscle or wind, or they arrived when at colder times (there were probably glaciers in the tops of the Cordillera Guanacaste in the last glacial maximum just 15,000 years ago) the upper elevation biota covered more continuous tracts at the lower elevations.

The foremost exception to this scenario is the Peninsula Santa Elena (today Sector Santa Elena and Sector Murcielago, bordered by Sector Marino [Area Marina Protegida]), about

30,000 hectares that was an island out in the Pacific 85 million years ago. This land mass joined the land bridge via tectonic affinities. It collided with the newly forming and slowly emerging Central America land mass 16 million years ago. Its insularity and its peculiar combination of serpentine soils (Sector Santa Elena, southern portion) and ancient marine soils (northern portion of Sector Murcielago) leads one to suspect the possibility that endemic species evolved *in situ* and have survived the invasion of their world by the continental biota.

IMAGES AND SPECIES ACCOUNTS

This book is, first and foremost, an effort to portray Lepidoptera as beautiful objects unto themselves. Since we are diurnal, color-vision mammals, the photographs we chose tend to depict species that are brightly colored. Moreover, by displaying them against a black background, we have further emphasized their vivid colors. As a result, the particular ledger of species we chose is biased in favor of aposematic species, those that warn the potential predator to stay away, and their mimics. The individual species accounts that follow the gallery of portraits make frequent reference to this aspect of their biology. The irony is that while it is easy to know that the bright colors and complex patterns and shapes play central roles in the lives of butterflies and moths vis-a-vis predators that hunt by sight, so little experimentation has been done that we do not know which are poisonous or distasteful and which are merely mimicking them. More confusing yet, we have only the slightest trace of an idea regarding which species are avoided because the predator has had an unpleasant experience with them or a look-alike, and which are avoided because the predator is genetically hard-wired to avoid that color or color pattern.

We have chosen just 100 species in the belief that less is more and because an individual book has its limits. The final portfolio was set, in part, by serendipity. We portray those species that we were able to photograph over the past twenty-eight years of the inventory and those species that were present when Jeffrey Miller visited the ACG. Some of the images have been edited. Scratched and torn wings and broken antennae have been digitally manipulated to present an illustration of a more perfect specimen, but no creative reconstruction was employed. We did not borrow colors, patterns, or body parts from an image of a different individual or fill in missing parts with freehand illustration.

The essays for each species that follow the gallery were written to reflect what Daniel Janzen or Winifred Hallwachs would say if an interested biologist or ecotourist walked up to one of them in the ACG with that butterfly, moth, or caterpillar in hand. Interim names, rather than formally described species names, have been used for some of the species. Each essay integrates a given species into a certain theme relevant to the natural history of that species, but does not pretend to offer an entire life history of that species. A more complete discussion will be possible only after the secrets of their lives are revealed to future generations of biologists.

A Gallery of Butterflies and Moths

1. *Moresa valkeri*

2. *Trosia* JANZEN01

3. *Automeris belti*

4. *Automeris io*

5. *Oryba achemenides*

6. *Oryba kadeni*

7. *Fountainea eurypyle*

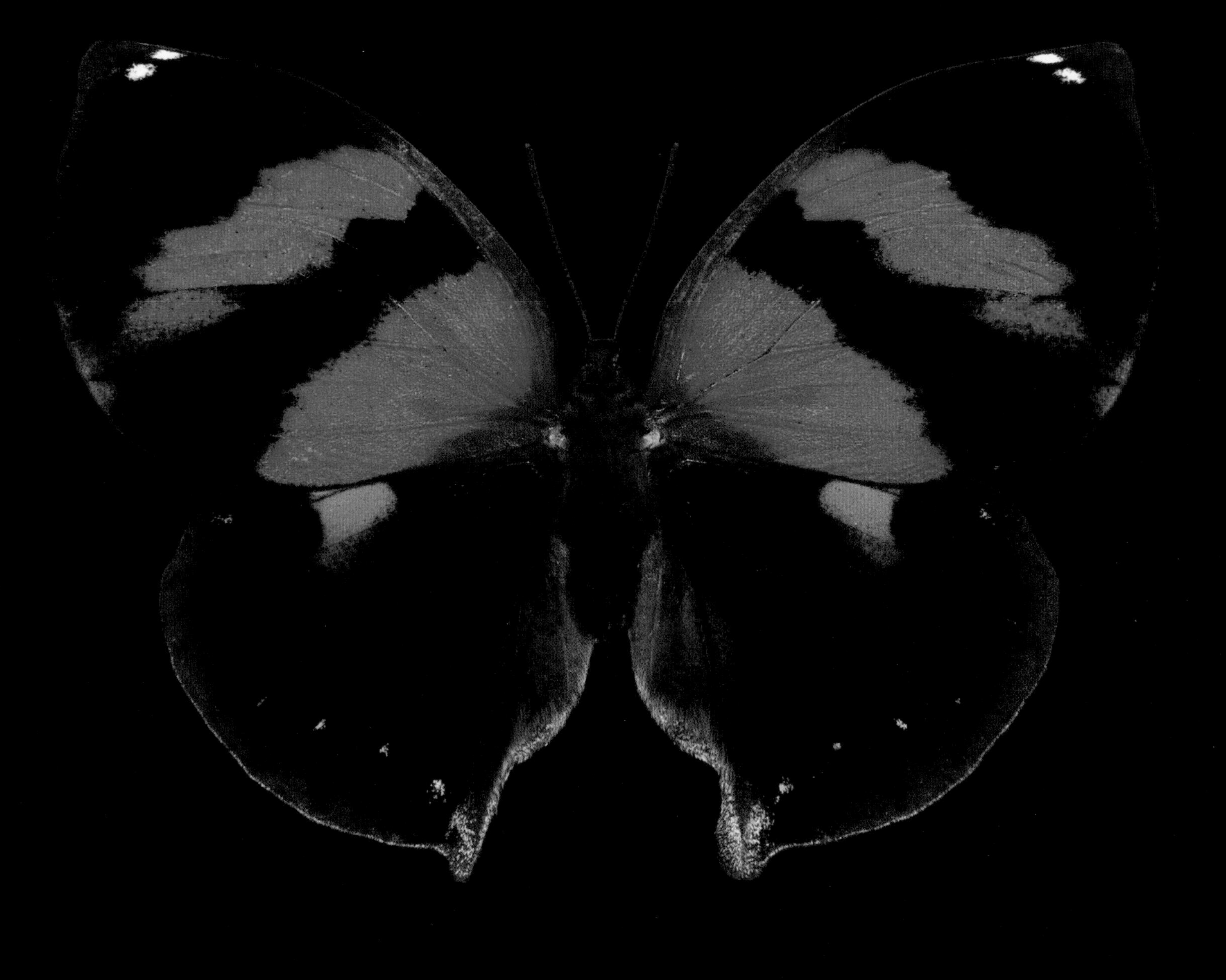

8. *Siderone marthesia*

9. *Astraptes* INGCUP

10. *Astraptes* YESENN

11. *Chrysoplectrum pervivax*

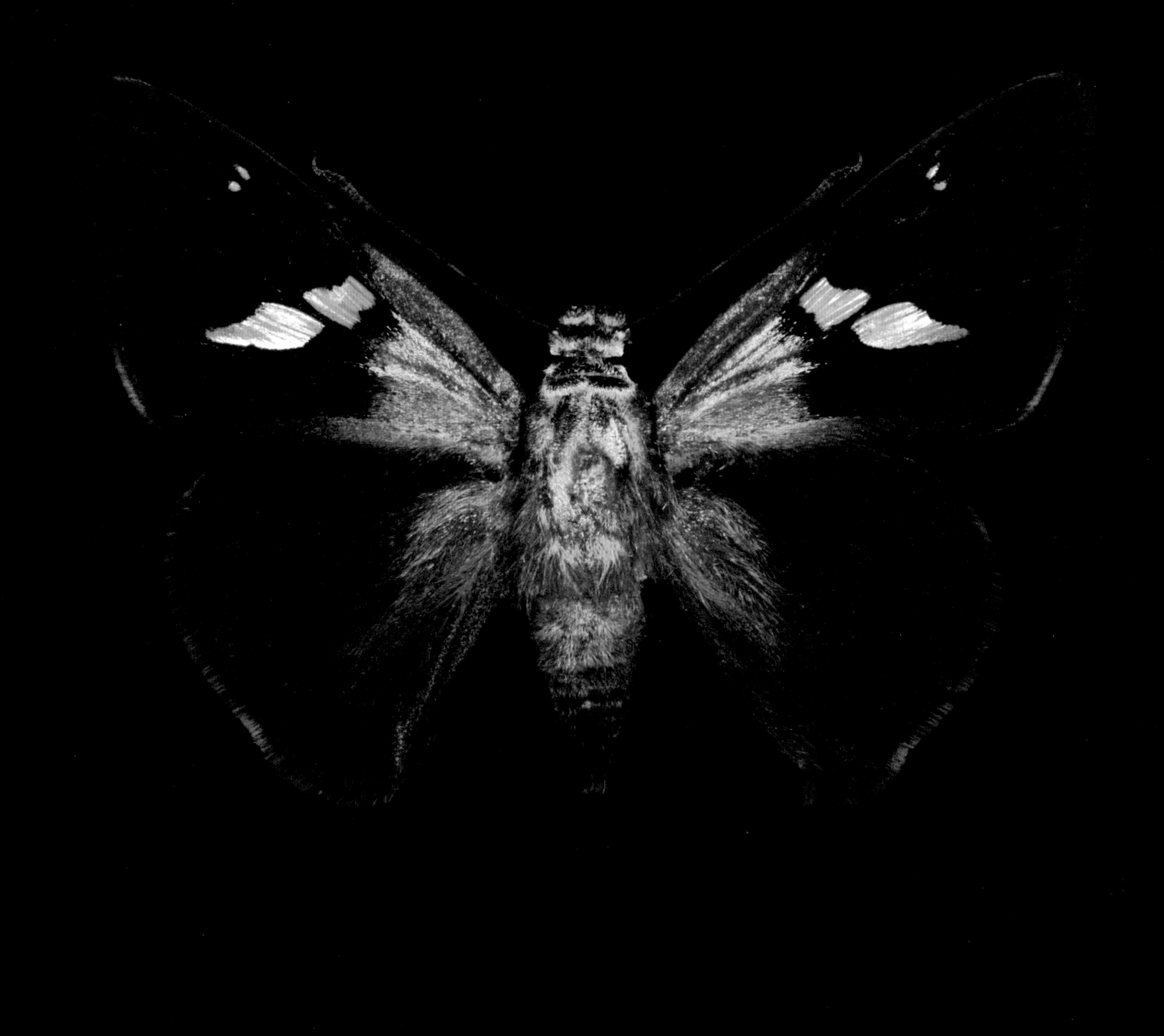

12. *Neoxeniades molion*

13. *Eueides procula*

14. *Tithorea tarricina*

15. *Xylophanes tyndarus*

16. *Adhemarius ypsilon*

17. *Parides iphidamas*

18. *Mimoides euryleon*

19. *Bardaxima perses*

20. *Calledema plusia*

21. *Zerene cesonia*

22. *Epia muscosa*

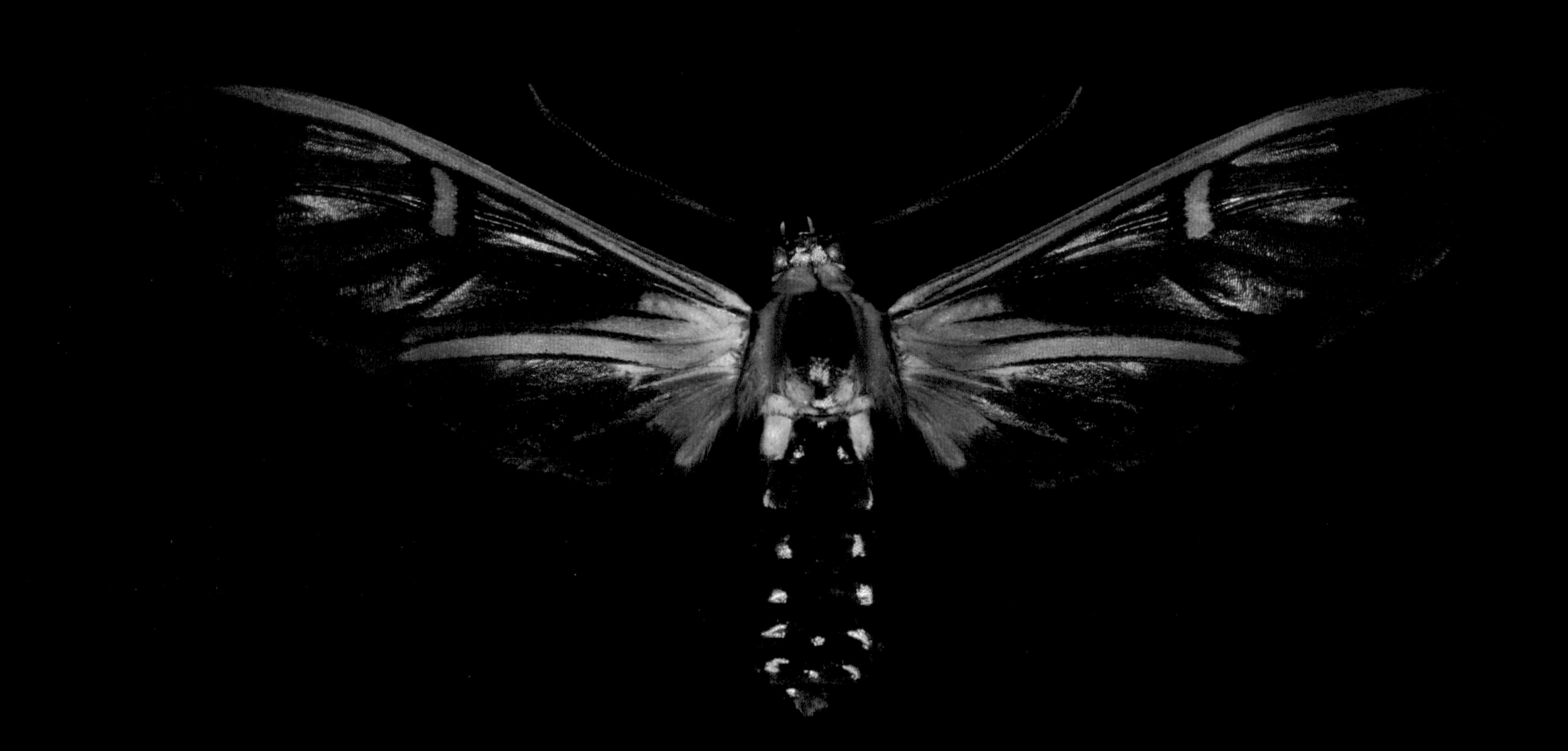

23. *Cosmosoma cingulatum*

24. *Cosmosoma teuthras*

25. *Historis odius*

26. *Anaea aidea*

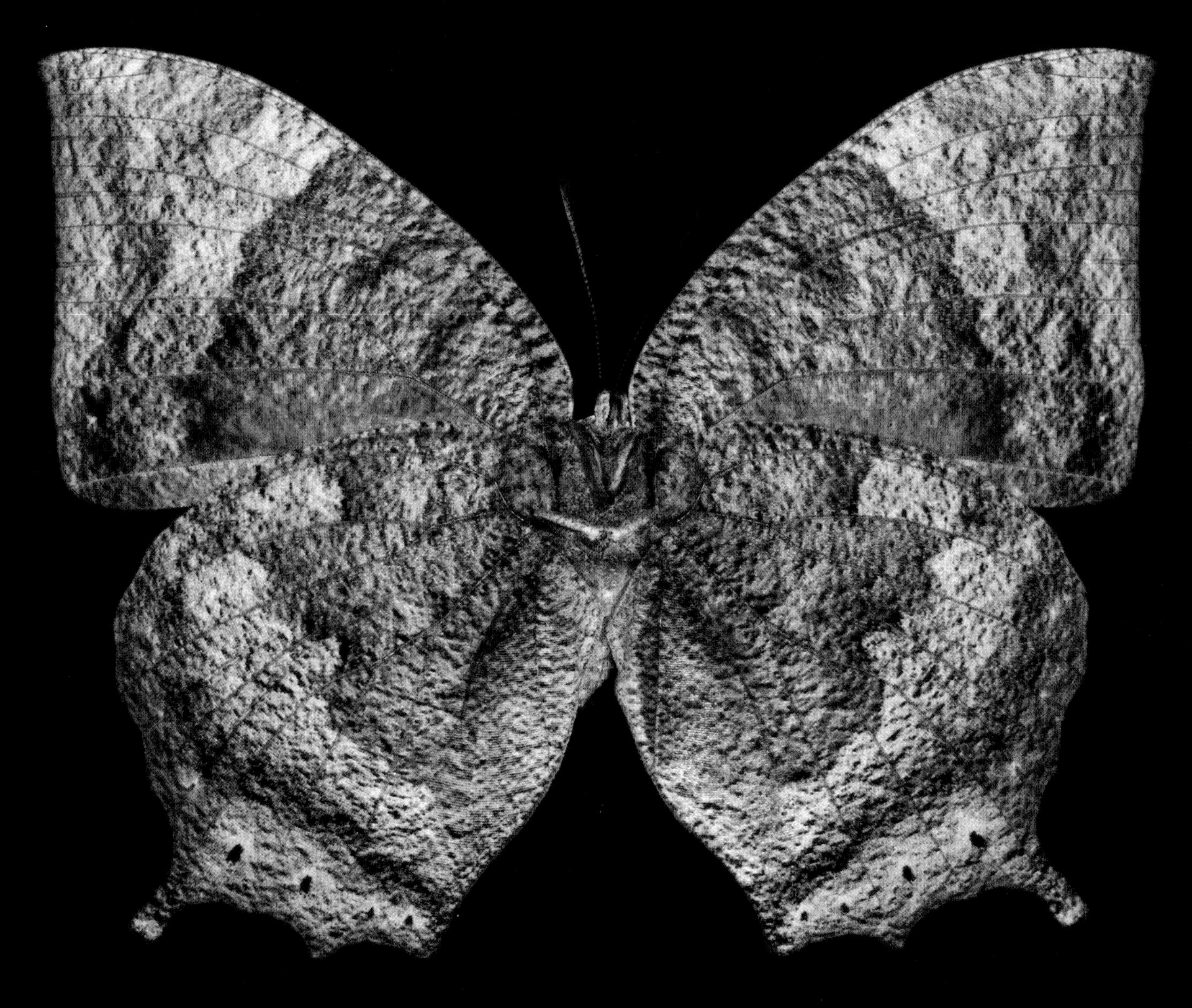

27. *Dysschema viuda*

28. *Danaus plexippus*

29. *Xylophanes porcus*

30. *Xylophanes crotonis*

31. *Siproeta epaphus*

32. *Adelpha basiloides*

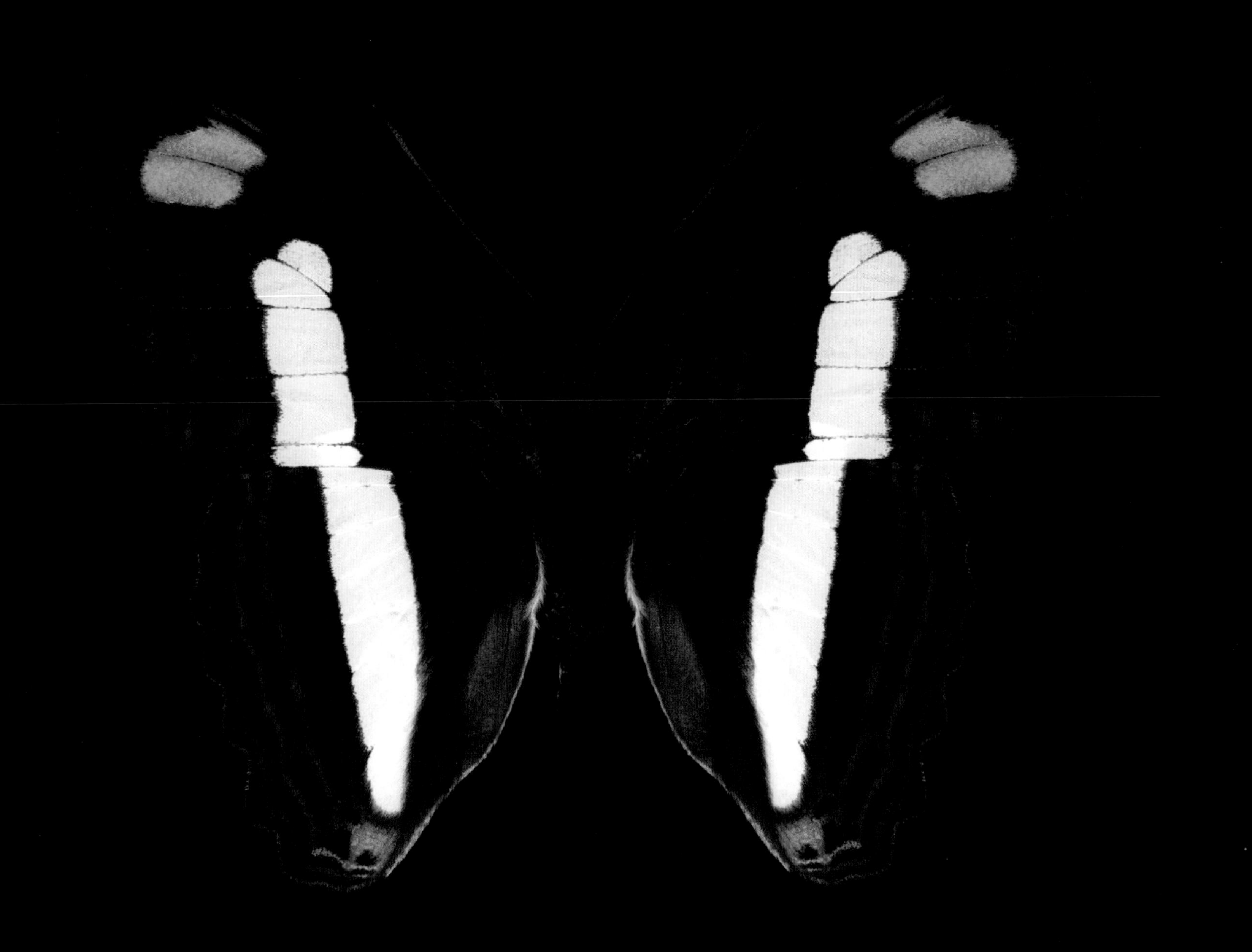

33. *Morpho theseus*

34. *Morpho amathonte*

35. *Opsiphanes cassina*

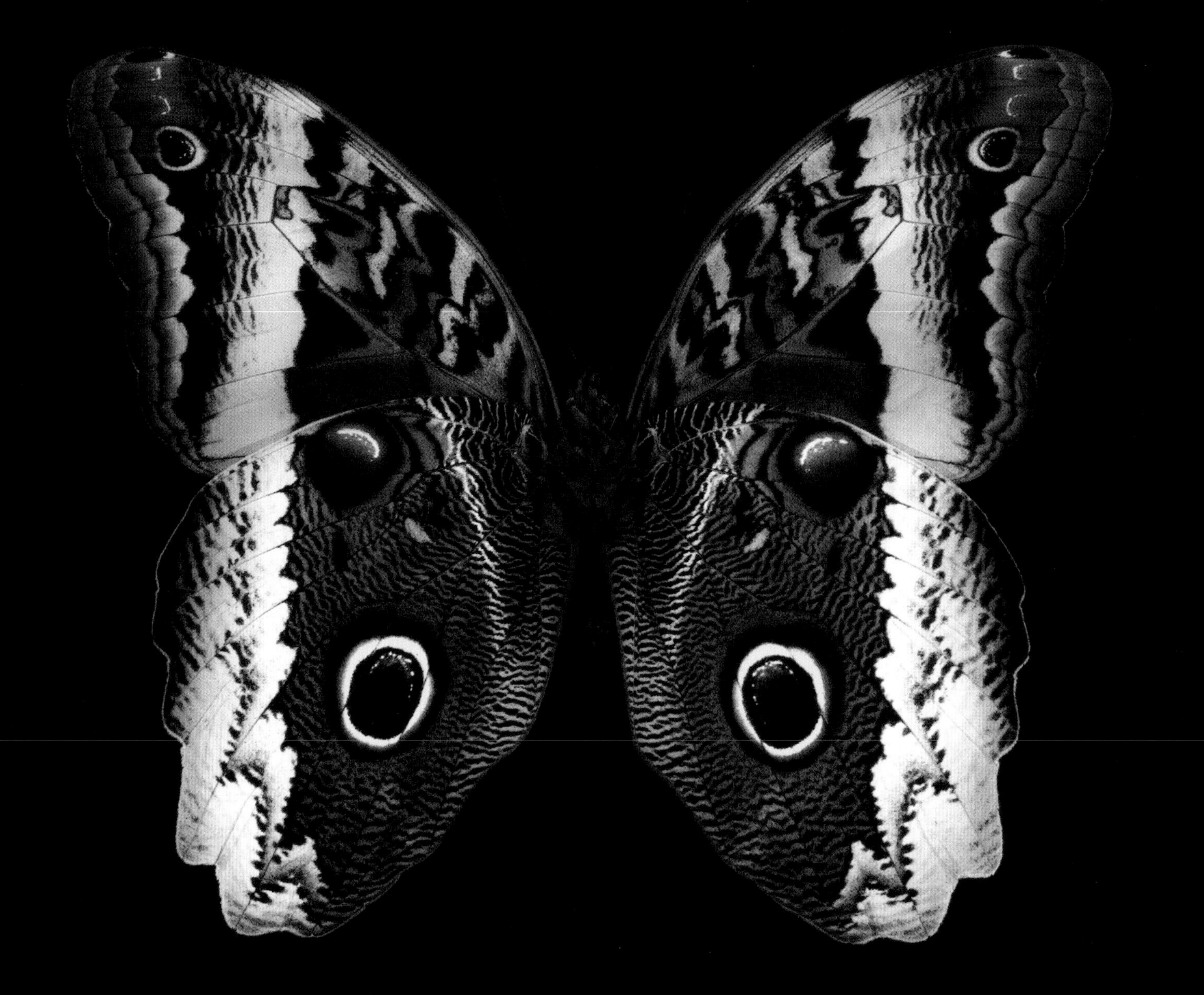

36. *Caligo atreus*

37. *Pentina flammans*

38. *Othorene verana*

39. *Memphis mora*

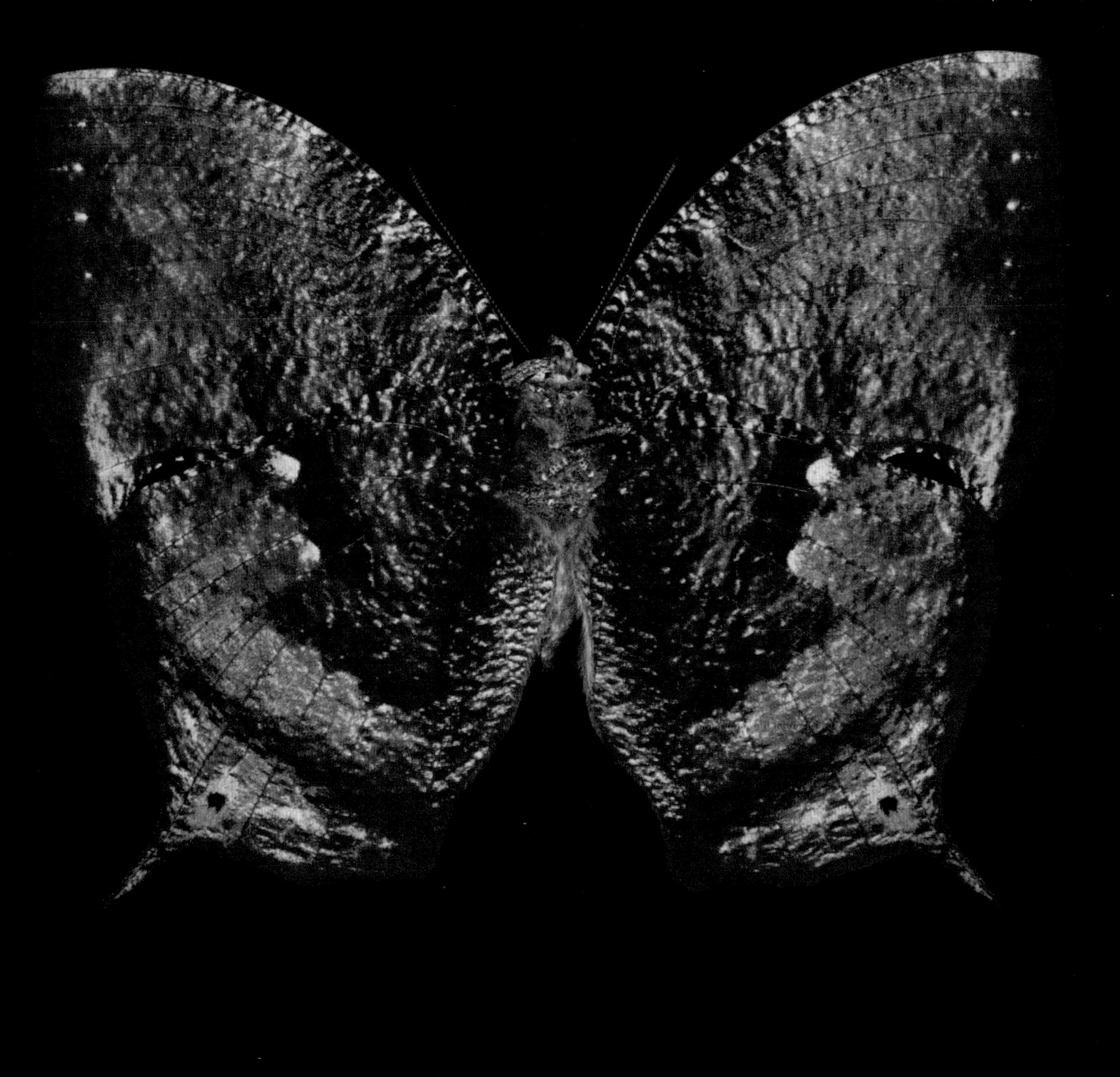

40. *Memphis proserpina*

41. *Mechanitis polymnia*

42. *Consul fabius*

43. *Arsenura drucei*

44. *Mimallo amilia*

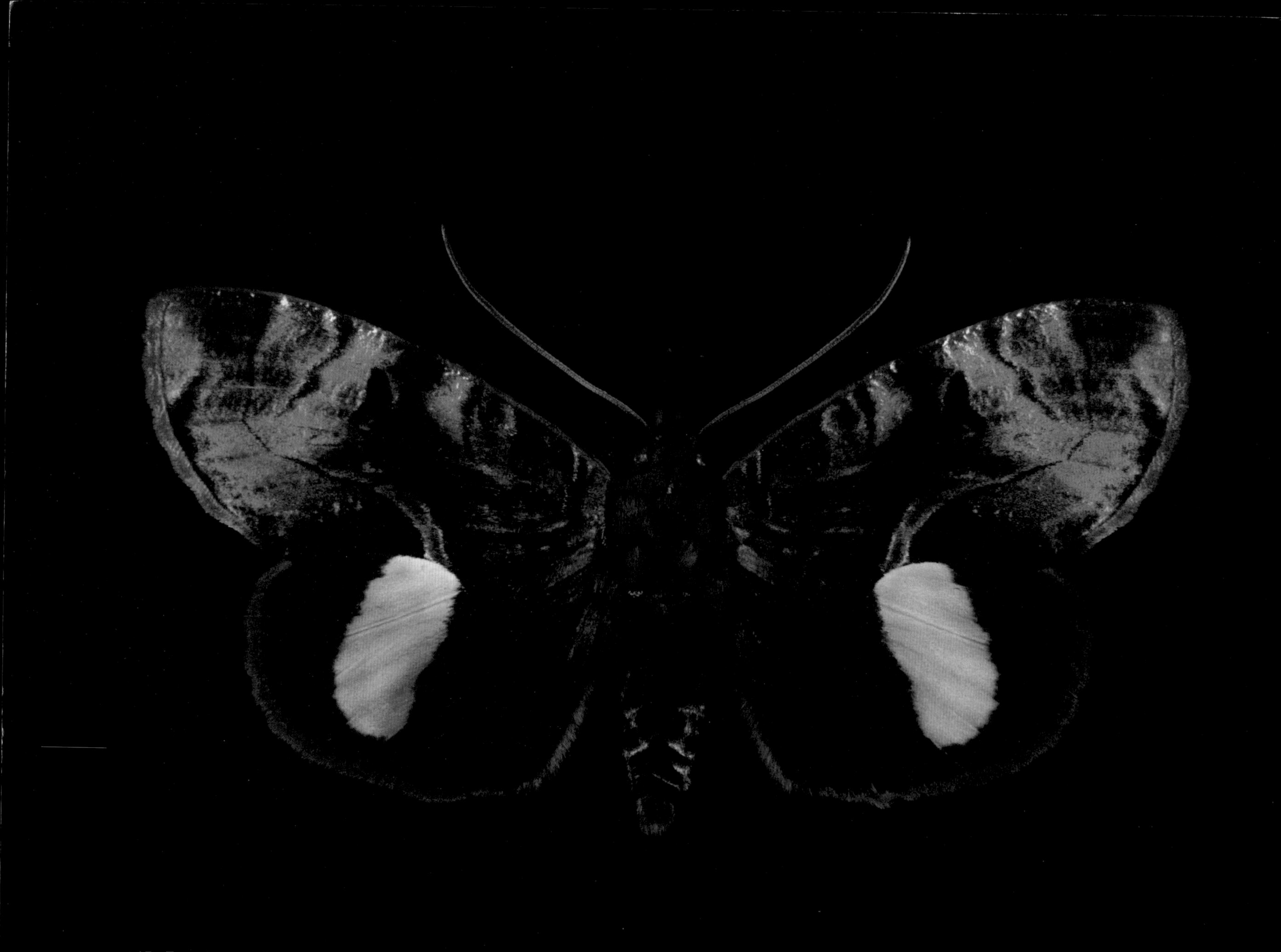

45. *Ferenta castula*

46. *Porphyrogenes* BURNS01

47. *Oxytenis modestia*

48. *Copaxa curvilinea*

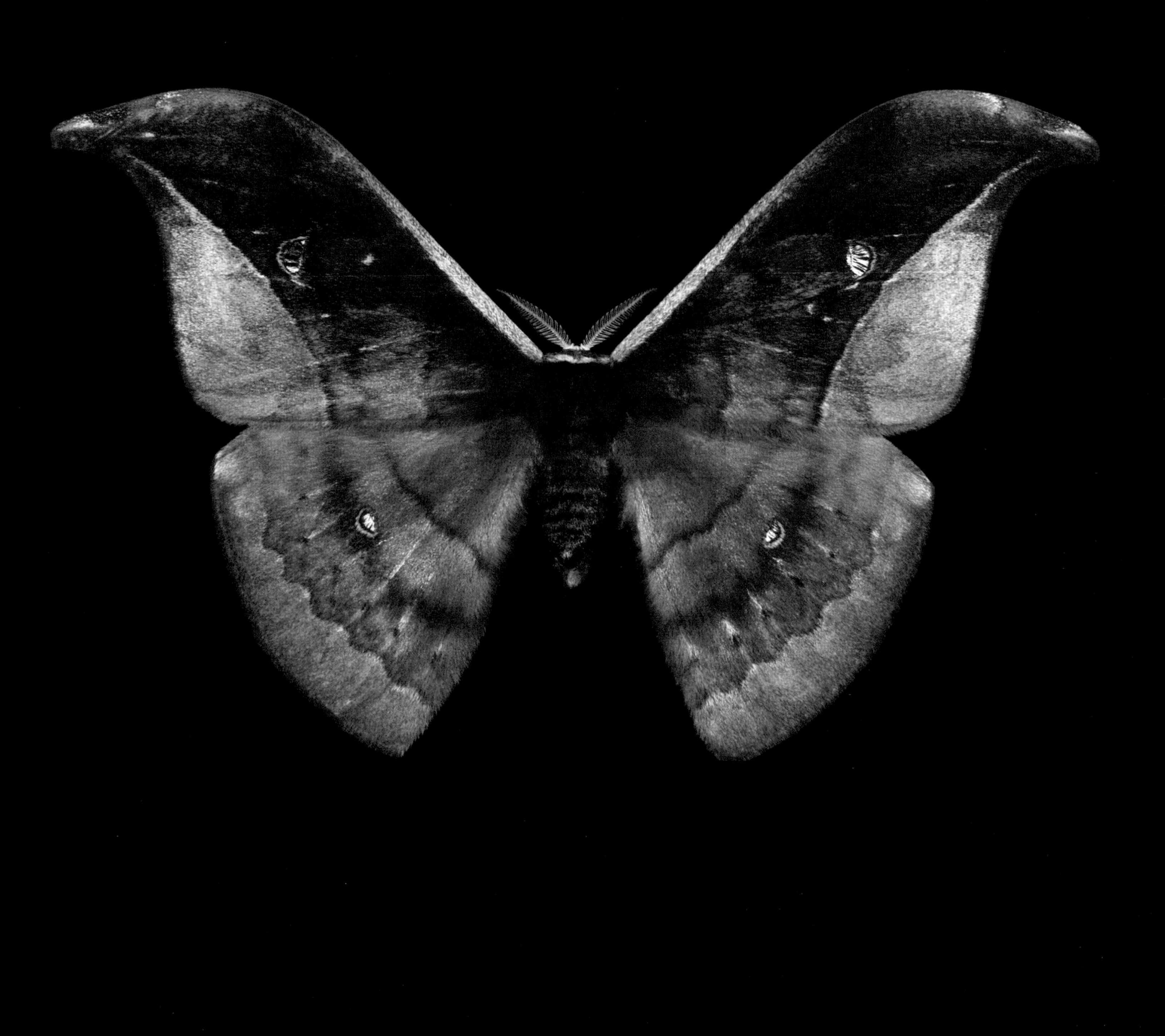

49. *Myscelia pattenia*

50. *Hamadryas amphinome*

51. *Zale peruncta*

52. *Letis mycerina*

53. *Hypercompe icasia*

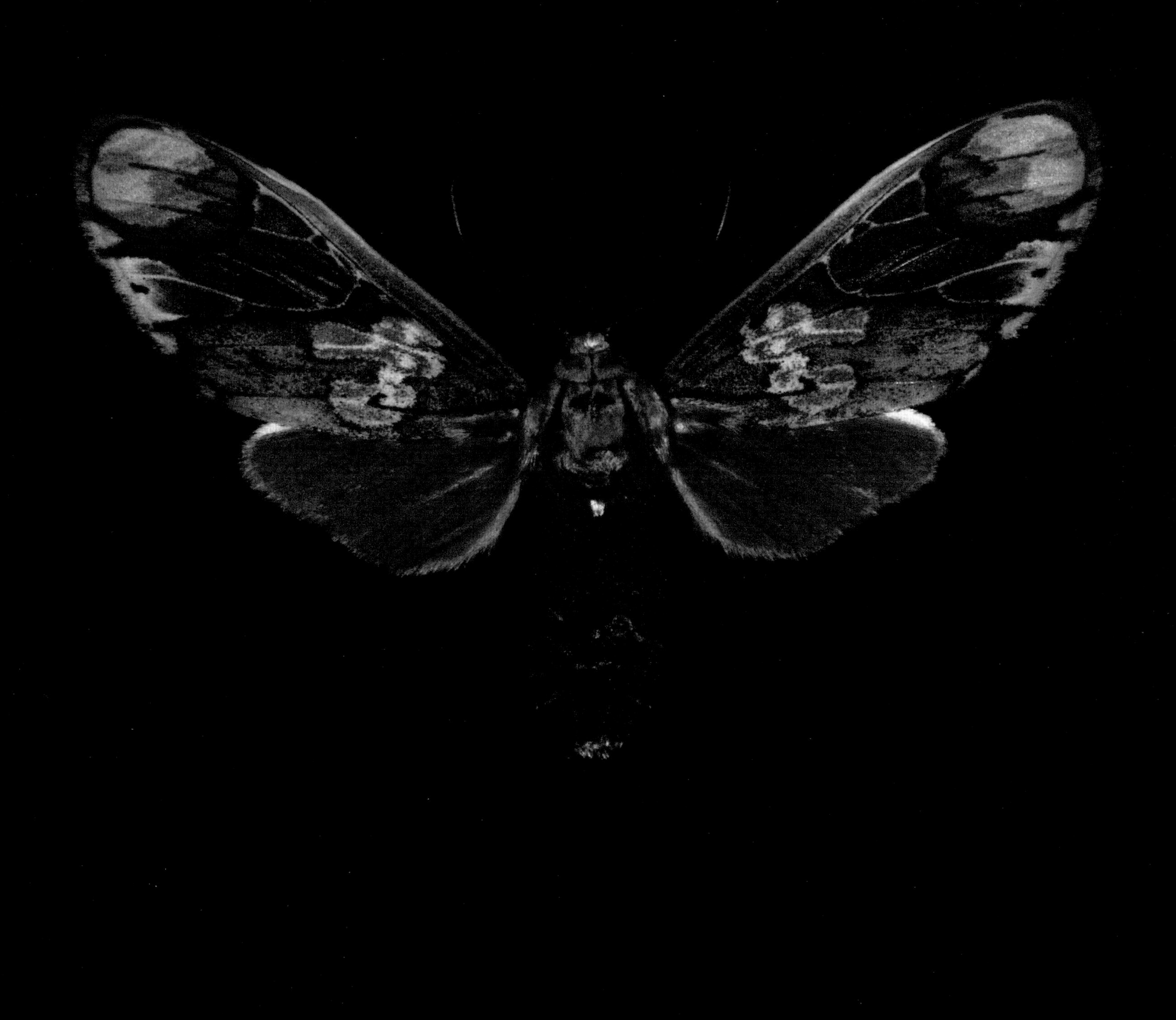

54. *Neonerita dorsipuncta*

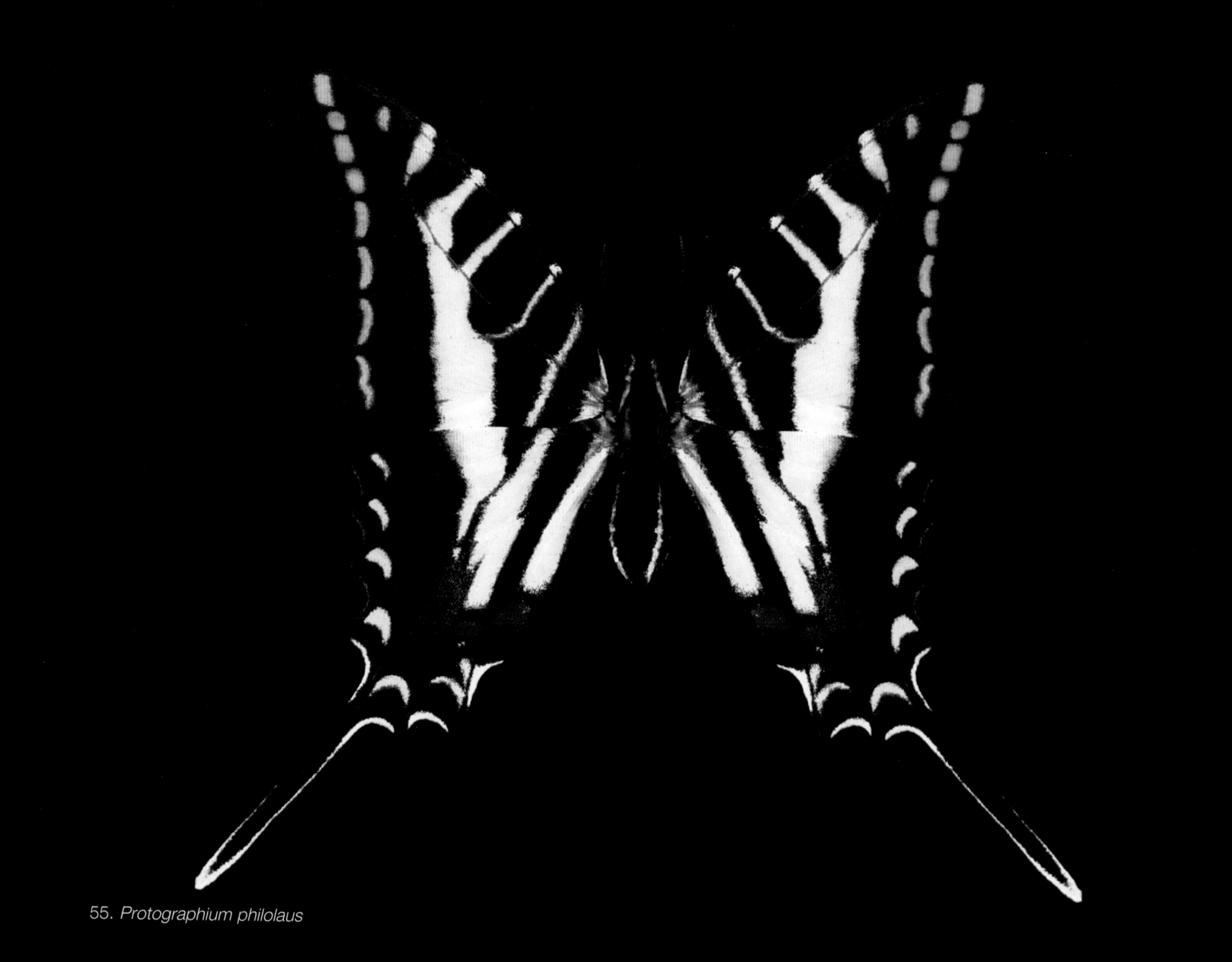

55. *Protographium philolaus*

56. *Protographium marchandi*

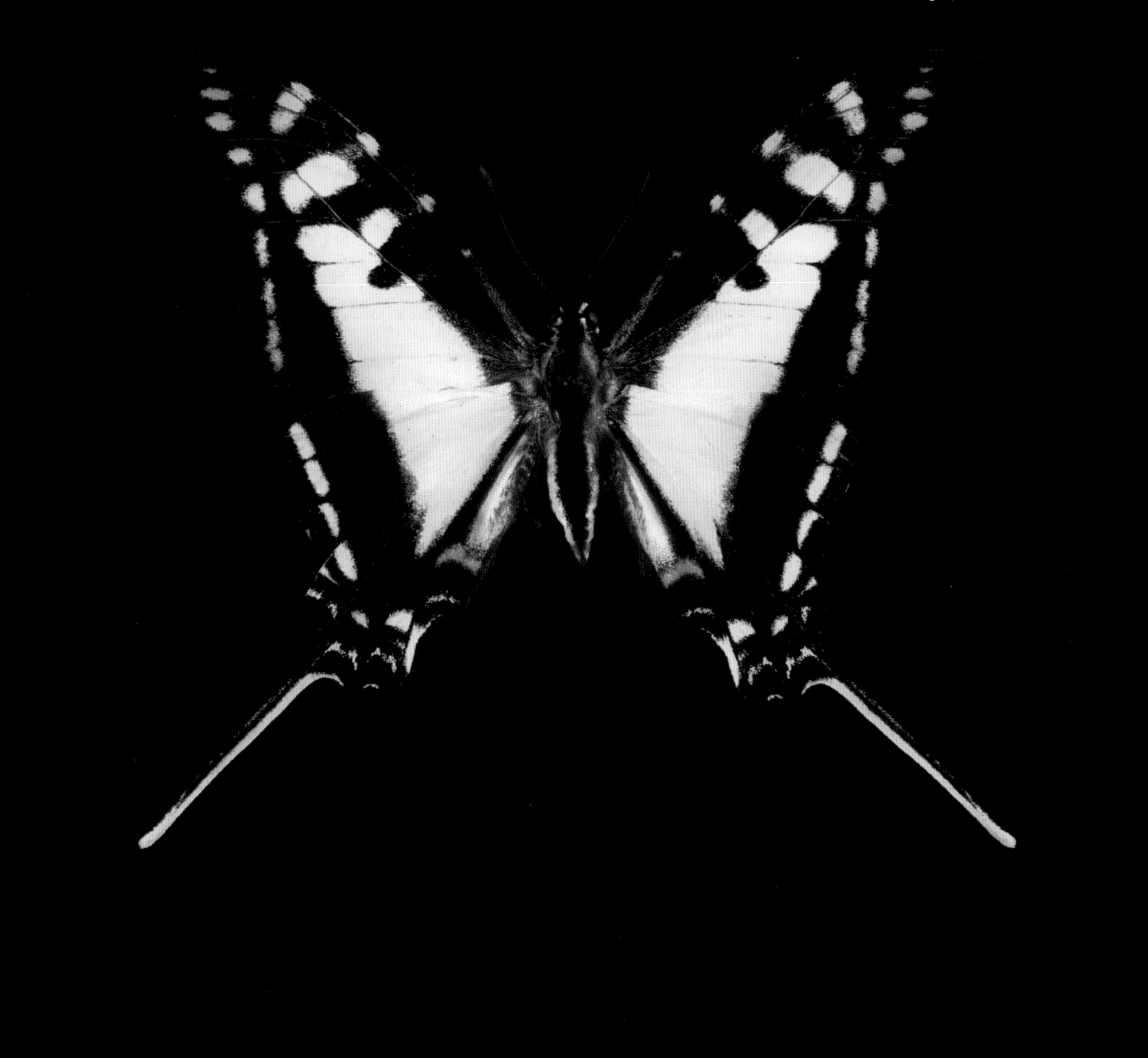

57. *Kloneus babayaga*

58. *Madoryx plutonius*

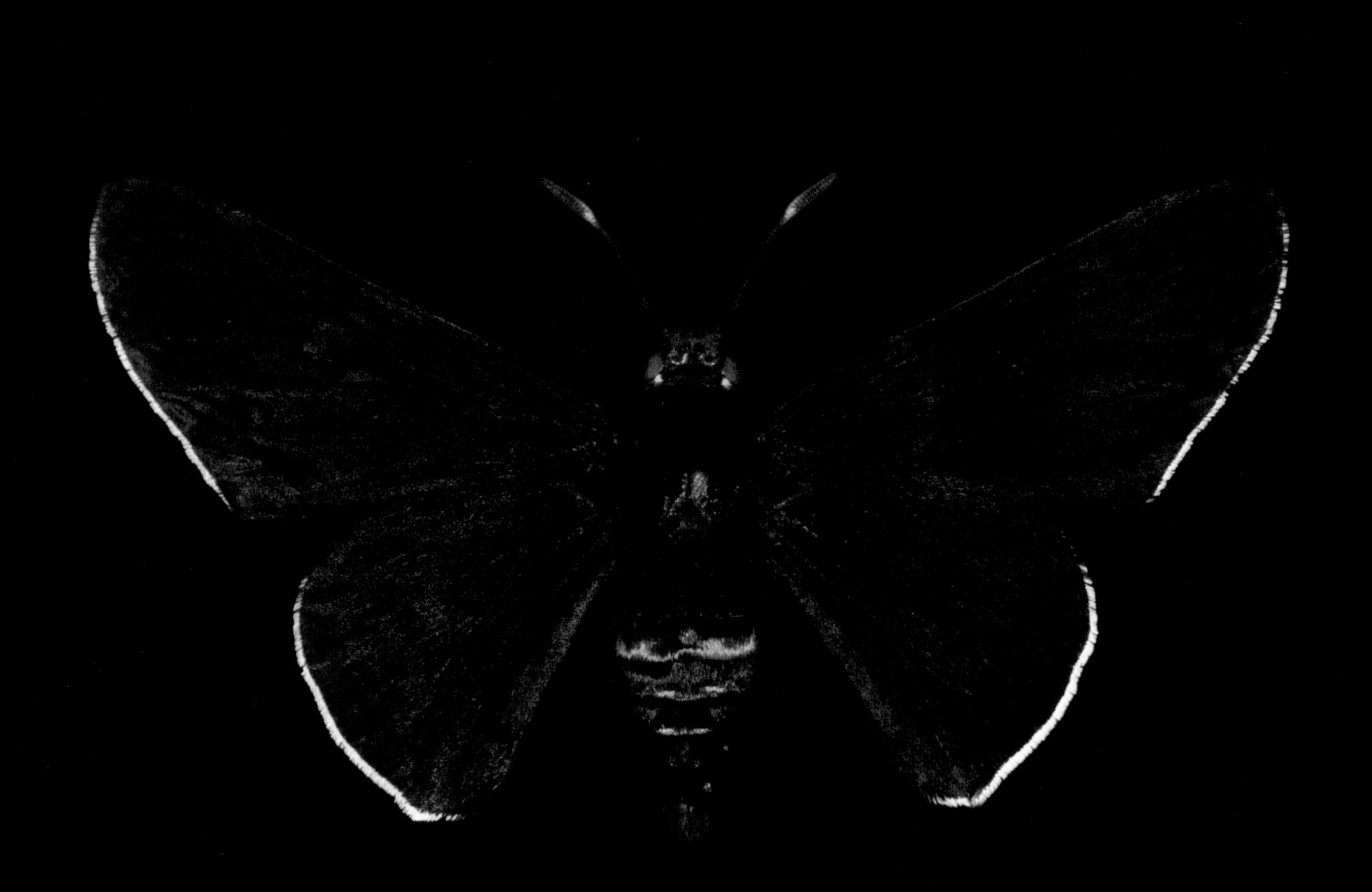

59. *Oxynetra hopfferi*

60. *Creonpyge creon*

61. *Bungalotis astylos*

62. *Bungalotis diophorus*

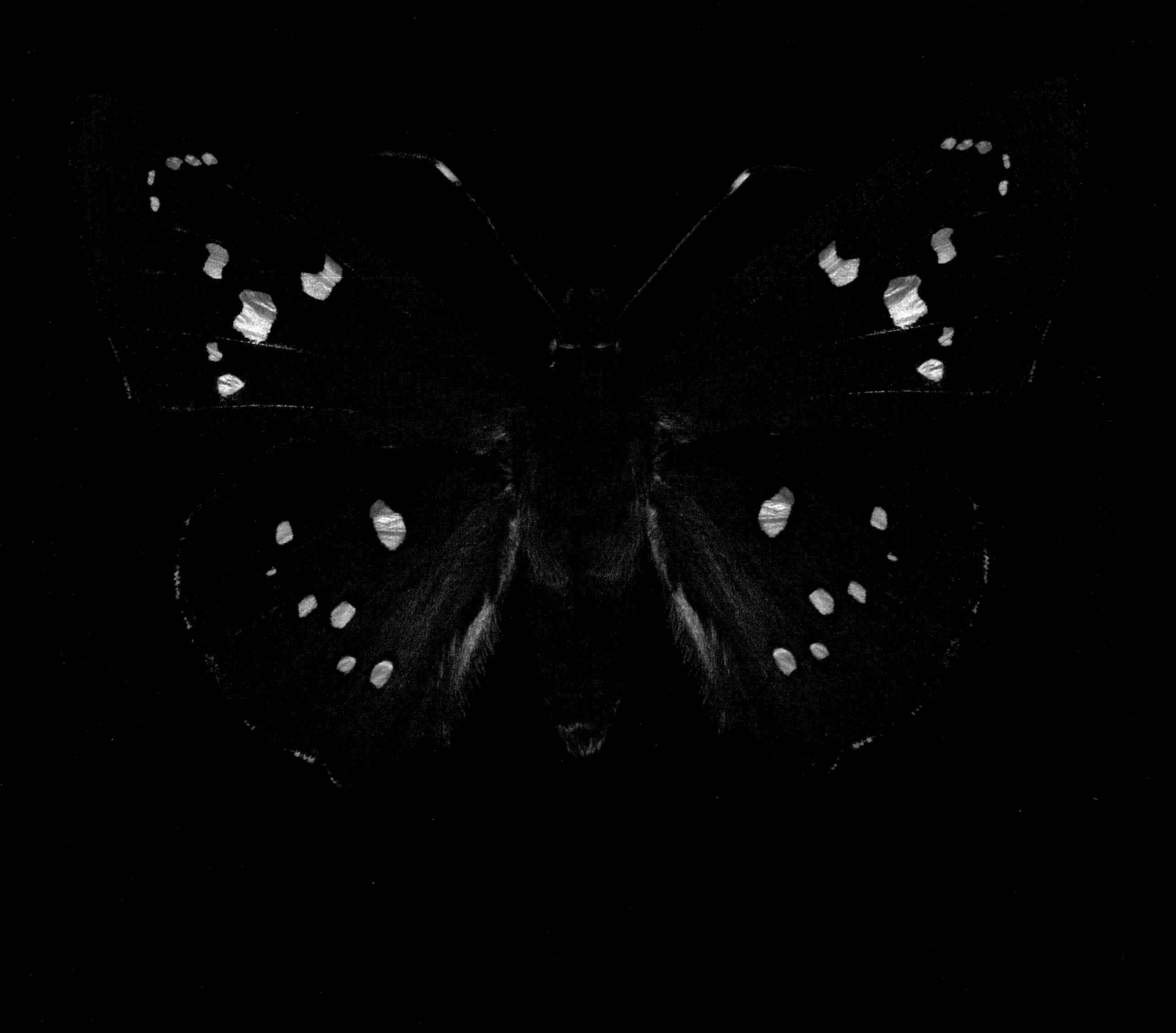

63. *Sarsina purpurascens*

64. *Rejectaria atrax*

65. *Erbessa salvini*

66. *Zunacetha annulata*

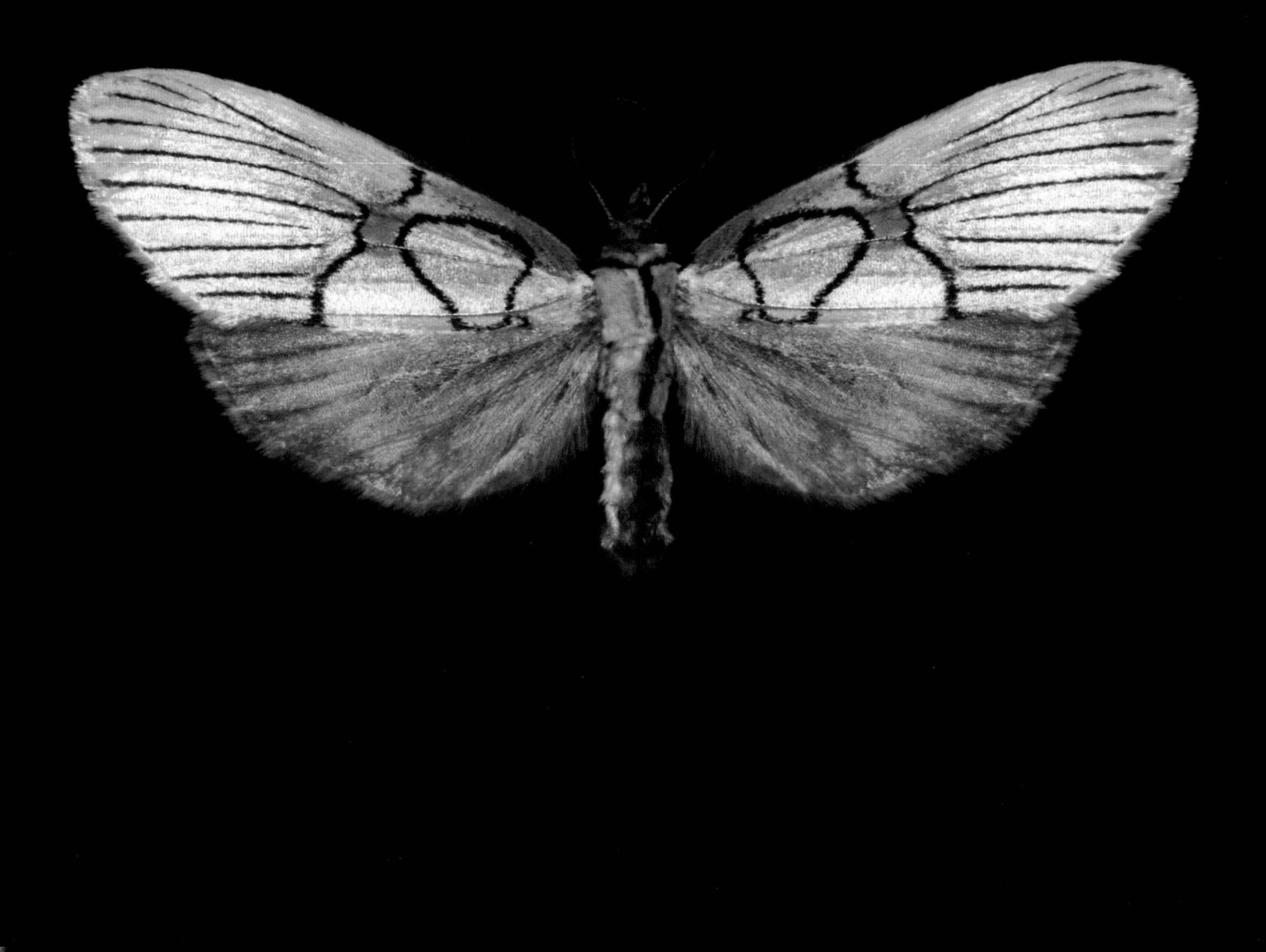

67. *Telemiades chrysorrhoea*

68. *Proteides mercurius*

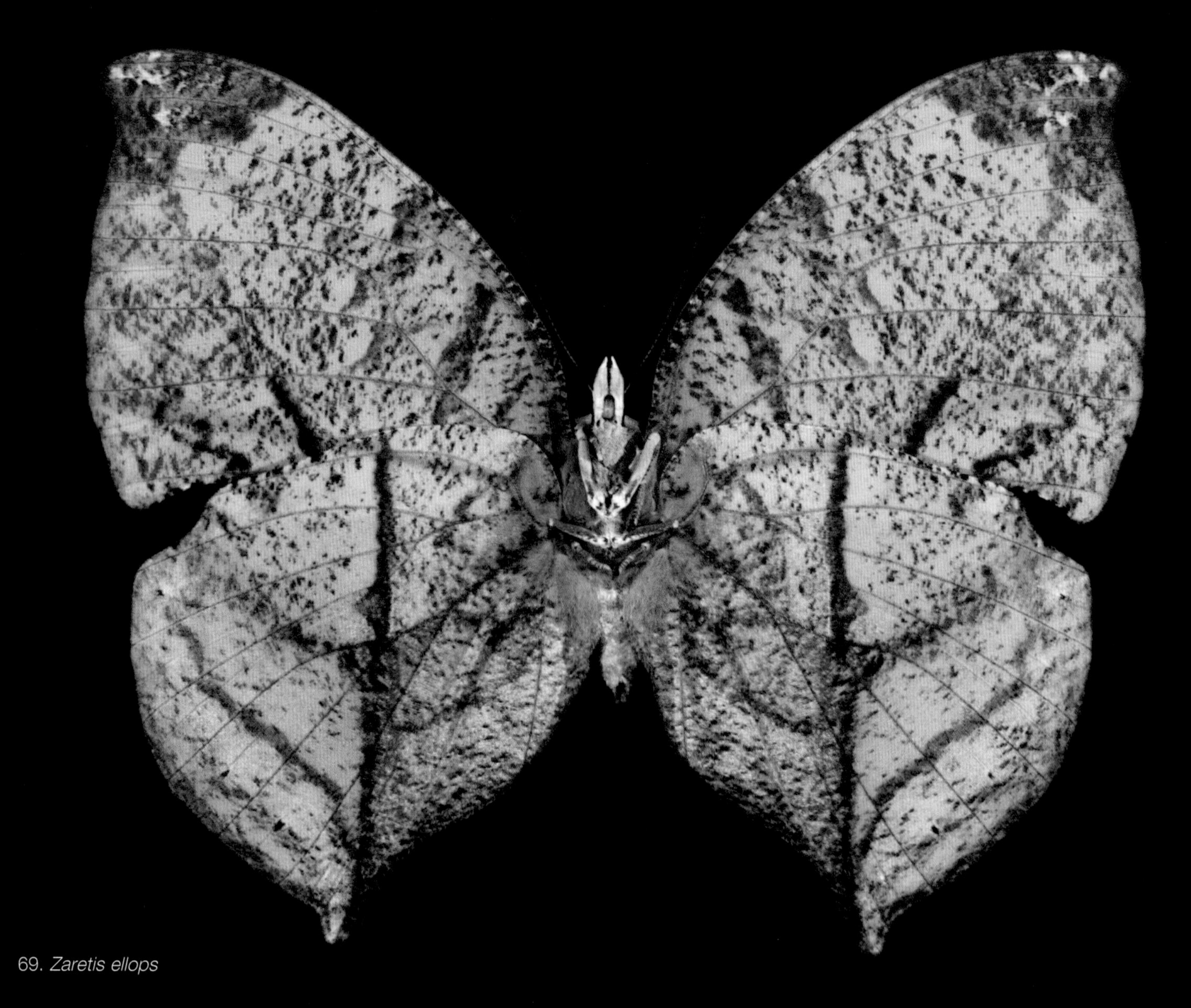

69. *Zaretis ellops*

70. *Zaretis itys*

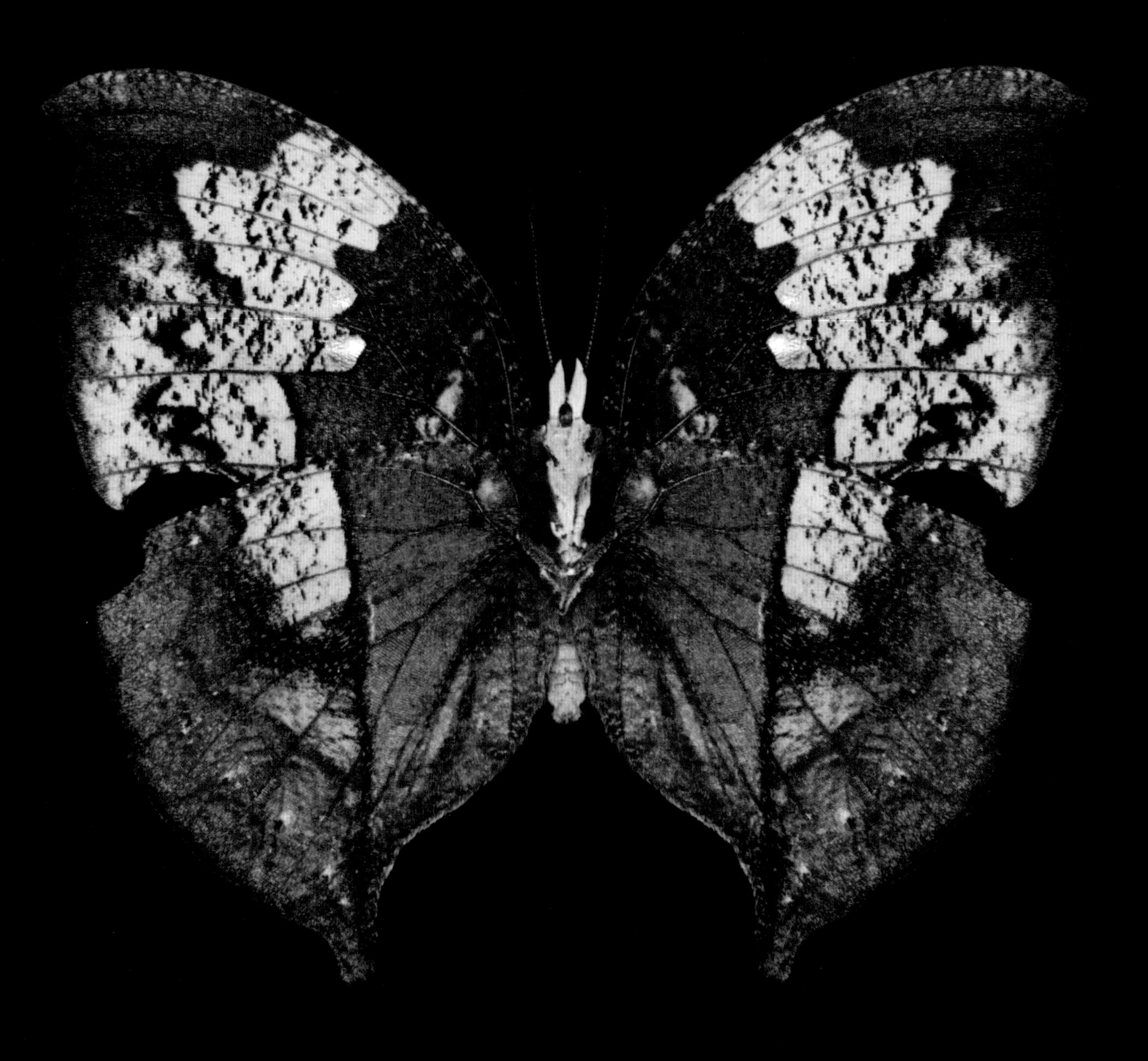

71. *Entheus matho*

72. *Dismorphia amphiona*

73. *Strophocerus thermesia*

74. *Gonodonta pyrgo*

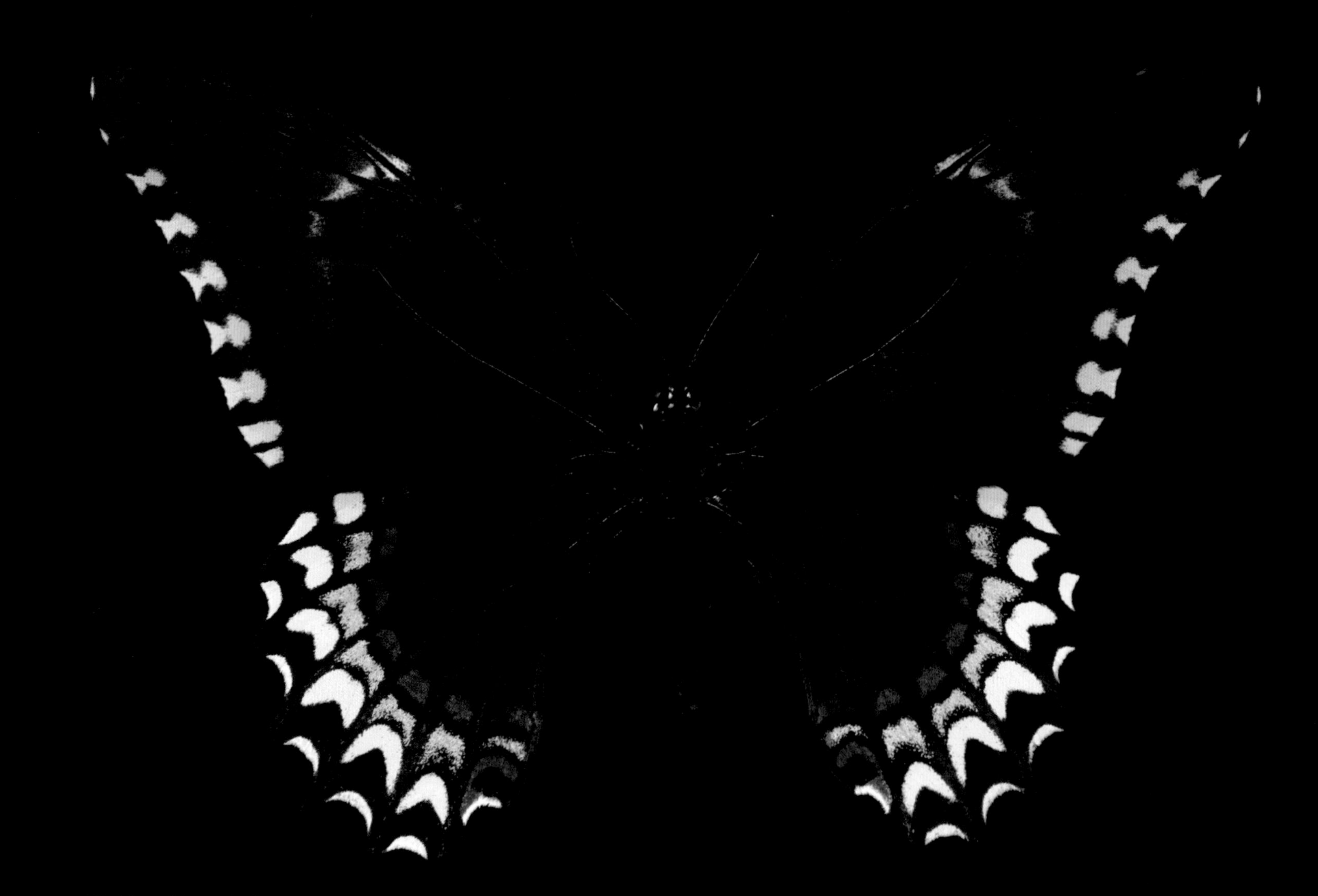

75. *Heraclides astyalus*

76. *Agrias amydon*

77. *Urbanus belli*

78. *Astraptes talus*

79. *Archaeoprepona demophon*

80. *Prepona laertes*

81. *Aphrissa statira*

82. *Phoebis philea*

83. *Aellopos ceculus*

84. *Enyo ocypete*

85. *Automeris metzli*

86. *Automeris phrynon*

87. *Cerura dandon*

88. *Cerura rarata*

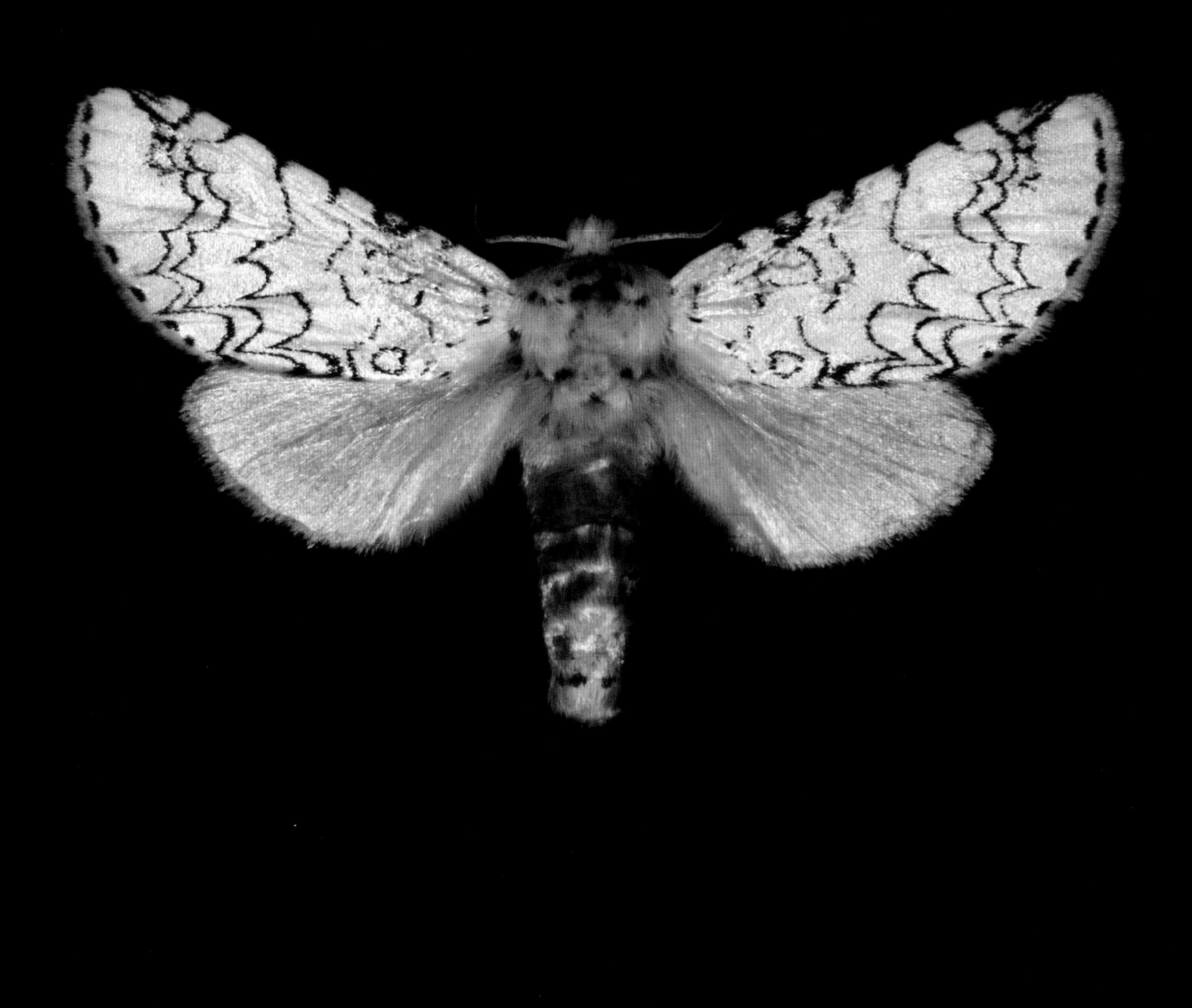

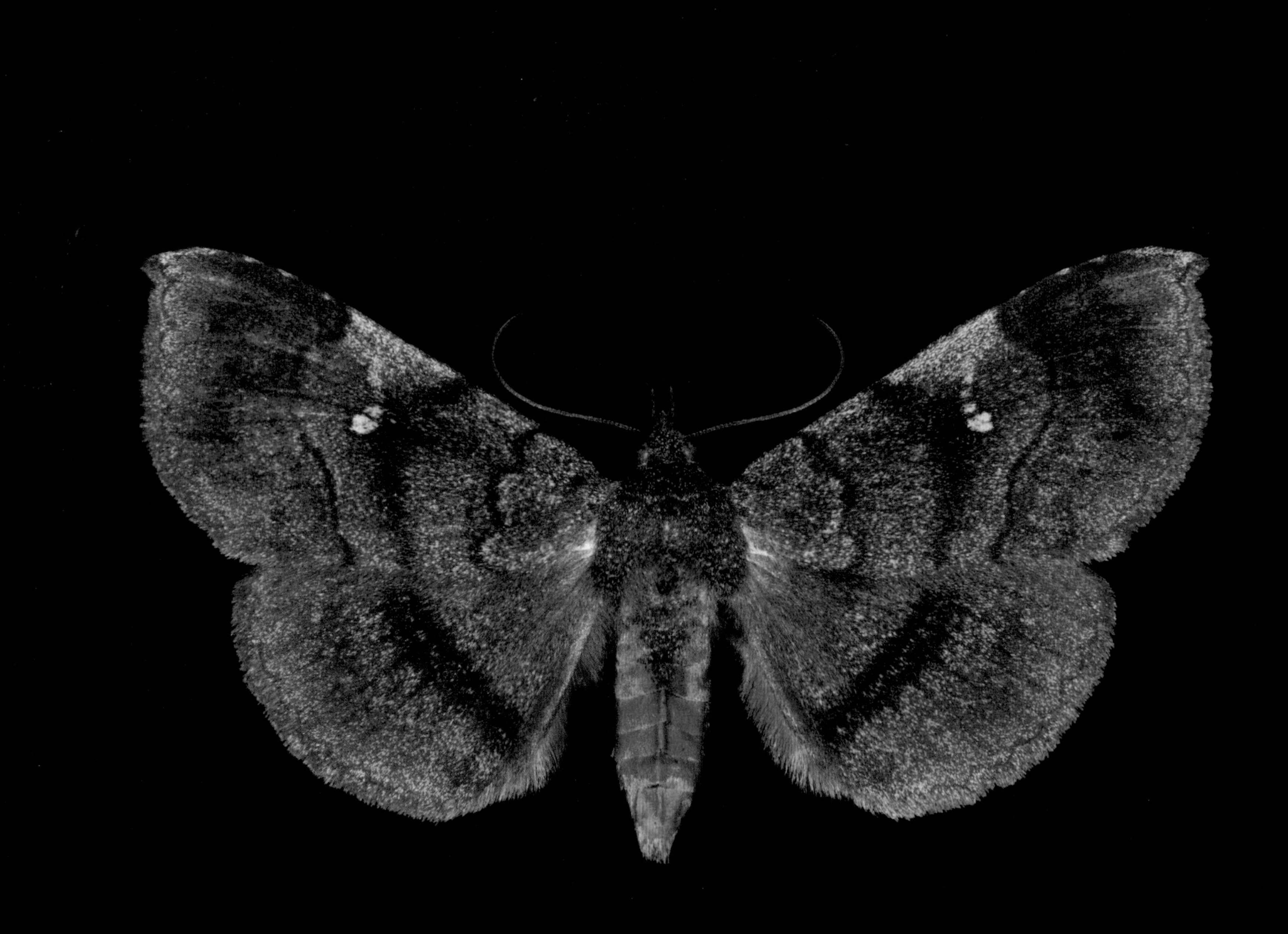

89. *Azeta rhodogaster*

90. *Pseudodirphia menander*

91. *Phocides nigrescens*

92. *Parelbella macleannani*

93. *Rhescyntis hippodamia*

94. *Rothschildia erycina*

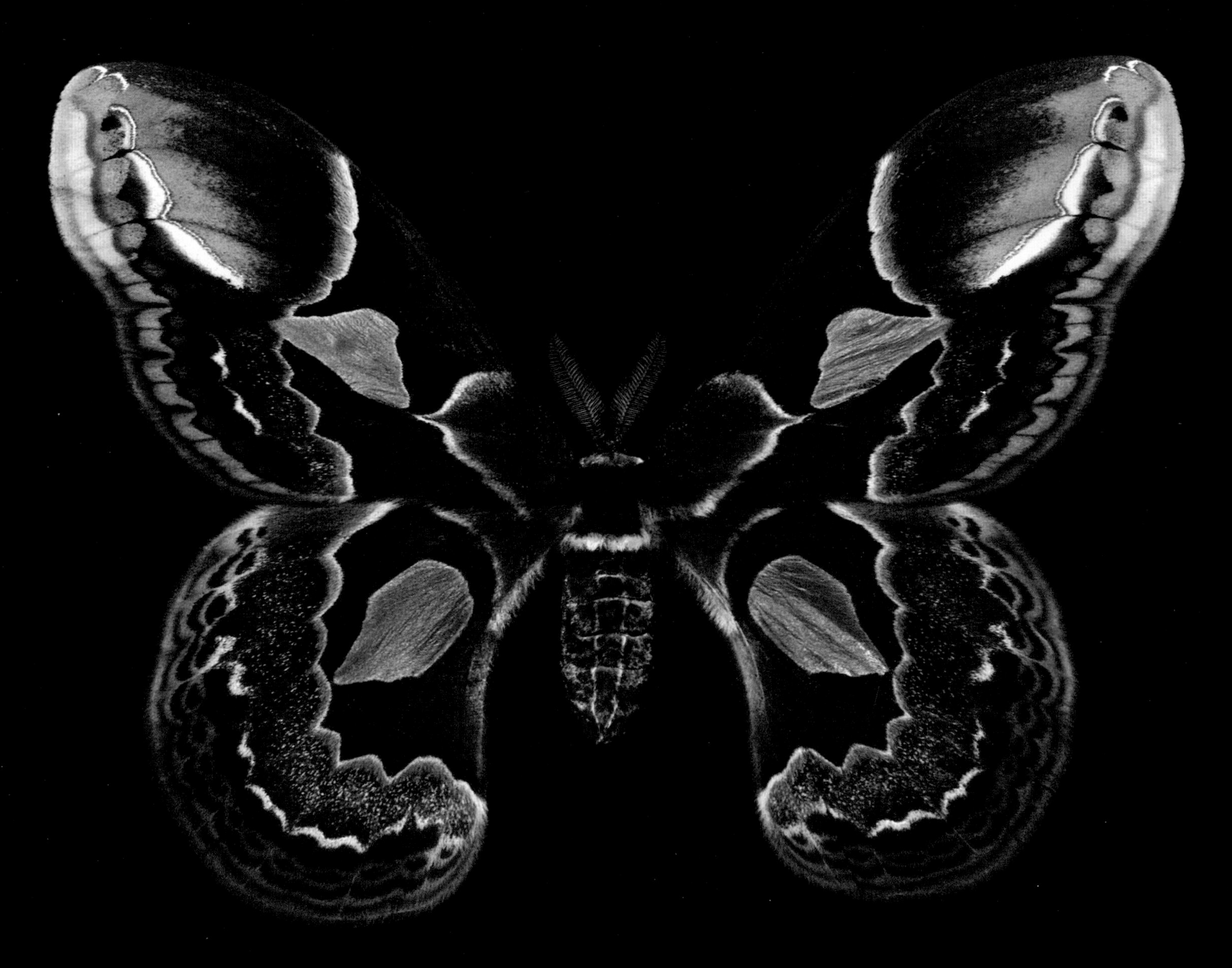

95. *Greta morgane*

96. *Mydromera notochloris*

97. *Cacostatia sapphira*

98. *Belemnia trotschi*

99. *Mesotaenia barnesi*

100. *Pierella helvetia*

Behavior, Ecology, and Caterpillar Images

1. *MORESA VALKERI* – NOTODONTIDAE

When sorting a huge sample of moths collected from light traps, one method is to look for a single color and pick out all the individuals sporting that color, then sort again. If the original sample consists of Notodontidae, an easy first round of sorting separates all the green moths, and in the ACG this will be species of *Kalkoma, Rosema,* or *Moresa.* The describer of the two latter genera was having fun with words, *Moresa* being an anagram of *Rosema.*

Moresa valkeri was originally described as *Rosema valkeri,* undoubtedly due to the members of both genera having leaf-green forewings, a most unusual color for a notodontid (and for most moths). It certainly was correct to split certain species into the new genus, *Moresa.* The distinctive, dorsally flattened *Rosema* caterpillars, which eat *Inga,* bear no resemblance to the cylindrical and tailed caterpillar of *M. valkeri.*

Variation in the wing markings in *M. valkeri* gives pause to the practice of using field markers for identification. Caterpillars reared at the same time from a single *Hymenaea courbaril* (Fabaceae) tree crown, their sole host, generated adults with forewings ranging from pure leaf green to having a huge, white, angled slash, and all grades in between. The individual we used in our portrait is between the extremes. In the case of *M. valkeri* the variation is most likely environmentally induced rather than genetically determined. As potential prey, the moths do not look exactly alike, thus no single search image works for any given predator. Aerial foraging birds have been seen to pluck fleeing adults out of the air and swallow them whole. The caterpillars patterned like dead twigs and damaged leaves appear to be edible as well.

ADULT VOUCHER: 84-SRNP-1134; JCM
CATERPILLAR VOUCHER: 02-SRNP-12636; DHJ

2. *TROSIA* JANZEN01 – MEGALOPYGIDAE

This brilliant white and light red-pink adult *Trosia* JANZEN01 probably lives a lie, though one never knows. Living on water and nutrients acquired as a caterpillar, as though it were a small saturniid moth, this nonfeeding female lives only a few days. (The similarly colored but smaller males with more pointed wings lack functional mouthparts for feeding as well.) She mates the first night and lays nearly all her eggs in the following two to four nights. Her brilliant wings are likely warning of a toxicity she does not possess. She is not a mimic of any arctiid or other highly toxic bright red-pink moth, but visually orienting predators likely avoid taking a chance. In contrast to the brightly colored and diurnally active moths and butterflies, in which the function of colors in courtship and in aposematism, or warning coloration, become interwoven to a degree almost impossible to separate, a moth like *T.* JANZEN01 does not suggest courtship value in its coloration.

The caterpillar of *T.* JANZEN01, like that of other mega-

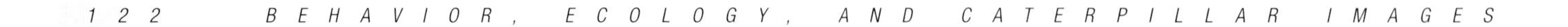

lopygids, is quite another story. It is densely and finely hairy and walks slowly and ostentatiously on all surfaces of the melastomataceous plants that it feeds on. It is dangerous, as any Costa Rican farming family knows. Present one to them and they recoil, call it a *tortulocuilo,* and most decidedly do not want to touch it. Experience is talking. Beneath the fine, fluffy hairs is a dense crop of sharp, stiff spines capable of delivering extraordinary pain-inducing chemicals. Merely touching these spines causes intense pain for hours, followed by a deep, long-lasting skin rash and then an eruption of sores. The pain and ensuing dermatitis are substantially worse than the pain inflicted by urticating hemileucine Saturniidae. There is no reason to think that the effect on a monkey or bird would be any different than on a human. Strangely enough, even this formidable defense is not foolproof. The squirrel cuckoos and trogons that regularly eat highly urticating hemileucine saturniids and feed them to their chicks do, on occasion, take megalopygid caterpillars as well. It goes to show that no caterpillar is totally free from predation.

ADULT VOUCHER: 00-SRNP-20956; JCM
CATERPILLAR VOUCHER: 05-SRNP-42079; JCM

3. *AUTOMERIS BELTI* – SATURNIIDAE

A glance through the color plates in Claude Lemaire's monumental monograph of the Neotropical hemileucine saturniids (Lemaire 2002) leaves the impression that if you have seen one *Automeris,* you have seen them all (with the exception of *Automeris phrynon,* 86). *Automeris belti* has the classic *Automeris* look—dull, dead-leaf forewings with glaring, huge false eyespots on the hindwings. It brings to mind an owl's eyes, but probably they are more generic false-eye superstimuli than selected to be an exact match of owl eyes. It makes one wonder if perhaps real owls' eyes are not themselves mimics of the more dangerous cat eyes, which are aposematic in their own right.

The males of *A. belti,* even siblings, display a wide range of dead-leaf colors, from dark brown to almost yellow-beige. We presume that this is yet another polyphenism or a polymorphism selected for by a vertebrate predator that forms a search image for its prey. However, one can wonder if *A. belti* is ever common enough to create a selective event of such magnitude as to embed the genes for such a wide range in color in the genome. The females are less variable. Their forewing colors, brown with purple and rust overtones, again raise the suspicion that simple polymorphic responses to predators may not be the entire story and that polymorphisms may have a range of effects beside predator evasion.

Although *A. belti* is a moth commonly found in rain-forest light traps, the bright yellow-green, almost fluorescent caterpillars are much less frequently encountered, at least down at the level of the lower foliage, the area most intensively searched by the *gusaneros.* It is likely, therefore, that most of the caterpillars typically forage higher in the canopy.

ADULT VOUCHER: 03-SRNP-12926.1; JCM
CATERPILLAR VOUCHER: 04-SRNP-60203; JCM

4. *AUTOMERIS IO* – SATURNIIDAE

Automeris io is really the icon for North American *Automeris* (though there are many other North American species in this genus; Lemaire 1988, 2002; Tuskes et al. 1996). It is *A. io* that causes so many North Americans' eyes to light up because they remember it as the bright yellow saturniid with big false eyespots on the hindwings. Somewhere, often on a gas station or restaurant window, they saw a yellow or reddish large moth and poked it, amazed when it popped the front wings forward and ostentatiously displayed the "eyes" on the hindwings. This behavior is a signature trait for the many tens of species of Neotropical *Automeris* and their close relatives, such as *Leucanella* and *Pseudautomeris.* This species is the most northern in the genus. As is the case with *Zerene cesonia* (#21), it is one of the Lepidoptera found in Minnesota that is also found in the ACG, the southernmost extent of its geographic distribution.

4

Color is a highly variable trait in *A. io.* Males are usually yellow with the eyespots layered on top, but occasional individuals within a clutch are a shade of pink, rust, and purple overall. Females wander all over the color map, from deep, reddish purple to a yellow almost as pure as that of the males. Sometimes the forewings lack color-coordination with the hindwings (see images on the inventory Web site at http://janzen.sas.upenn.edu). Such color inconstancy in ACG saturniids is often based on the season of eclosion and therefore likely related to temperature. Color inconstancy is also clearly a color polymorphism driven by search-image-forming predators. However, the range and mix of colors in *A. io* may be determined by an undetected environmental factor or selective process. Although an alternative explanation for color is that it is a random event, at least the variation in *A. io* is relatively symmetrical. Each of the forewings is usually the same color.

The bright green, highly urticating caterpillars of *A. io,* with their bright red and white lateral racing stripe, are also a part of childhood memories. Learning that this particular "woolly" or spiny caterpillar is painful when picked up is a memorable experience. The slow, diurnal, and ostentatious caterpillars feed on a long list of species of Fabaceae growing in highly disturbed habitats. They also eat species in a few other families of plants. In the ACG the caterpillars are very difficult to distinguish from those of *A. pallidior,* but the saving grace in the identification of many specimens is that *A. io* occupies the dry forest and *A. pallidior* occupies the interface between dry and rain forest. Yet another caterpillar look-alike, *A. celata,* occupies the ACG rain forest.

ADULT VOUCHER: 94-SRNP-9446.6; JCM
CATERPILLAR VOUCHER: 82-SRNP-850; DHJ

5. *ORYBA ACHEMENIDES* – SPHINGIDAE

In striking contrast to the caterpillar of *Oryba kadeni* (#6), which feeds on the leaves of a rubiaceous treelet commonly found in the rain-forest understory, *Cosmibuena macrocarpa,* the caterpillar of *Oryba achemenides* is without doubt the most localized sphingid caterpillar known from the entire ACG. To date, all eighty-six caterpillars of *O. achemenides* have been collected from one single, enormous vine of *Uncaria tomentosa* (Rubiaceae) growing in intermediate-elevation rain forest on the north foothills of Volcán Orosí. But the vine, too, is very rare. We have located only three individuals of this species of plant in the entire ACG. Two of the plants have not supported caterpillars. Although there must be others, it is nonetheless an extremely rare species in the ACG.

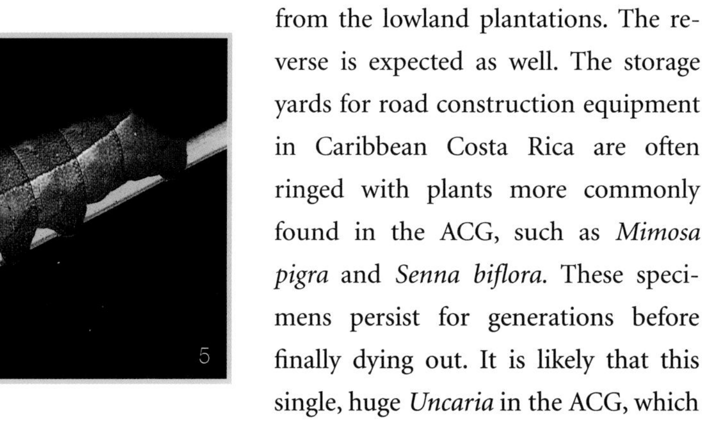

In fact, the food plant is so rare we wondered if it might be introduced from a distant, lowland rain forest, and therefore not really an ACG plant. How could this happen? The one individual food plant that has produced all of the known *O. achemenides* caterpillars in the ACG is rooted in the disturbed margin between the Del Oro orange plantations and adjacent uncleared rain forest. One of the other two individuals is also at an intermediate elevation between the plantations and the forest. The plantations were established through clearing and leveling of large tracts with heavy machinery. This machinery works all over Costa Rica, and specifically in the banana plantations in the Caribbean lowlands, where *U. tomentosa* occurs more frequently. When the machinery arrived at Del Oro for the preparation of the orchards, it would have been caked with mud from a previous job location. That mud could have contained seeds from the lowland plantations. The reverse is expected as well. The storage yards for road construction equipment in Caribbean Costa Rica are often ringed with plants more commonly found in the ACG, such as *Mimosa pigra* and *Senna biflora.* These specimens persist for generations before finally dying out. It is likely that this single, huge *Uncaria* in the ACG, which could have easily grown to its size since the 1990 establishment of the orange orchards, is an introduced plant. It is likely not just a coincidence that the adult of the single rain-forest *O. achemenides* collected in ACG in 1995 was from a light trap only three kilometers from this *Uncaria.*

Oryba kadeni and *O. achemenides* are so similar as adults that it would never be doubted that they belong in the same genus, and indeed their early instars and pupae are similar as well. However, their last instars (shown here) bear no resemblance to one another in color, pattern, morphology, or behavior.

ADULT VOUCHER: 04-SRNP-22749; JCM
CATERPILLAR VOUCHER: 04-SRNP-22759; JCM

6. *ORYBA KADENI* – SPHINGIDAE

The gorgeous greens and yellows of *Oryba kadeni,* a color less commonly encountered in adult Lepidoptera but so common for caterpillars, render this moth essentially invisible while it rests during the day. We know nothing of its flower-visitation behavior, but for a sphingid it has a relatively short tongue, suggesting it visits open flowers, such as those of the many rain-forest species of *Inga* (Fabaceae). Its very heavy and stocky body suggests that its flight may be much like that of a hummingbird, and a 2.5 gram female *O. kadeni* is about the same as the smallest ACG rain-forest hummingbird.

The first *O. kadeni* encountered by the inventory project came to a light trap set deep in the forest understory at Estación Pitilla. Usually, most species of sphingids appear readily at lights exposed on hilltops, in full view of the forest canopy. The resulting suspicion that *O. kadeni* is a species found deep in the forest understory rather than one that finds food among plants colonizing old fields and pastures, or early secondary succession, was confirmed when a *gusanero* found eggs on *Cosmibuena macrocarpa,* a plant that grows along disturbed edges of relatively intact rain forest.

Another oddity in the appearance of *O. kadeni* at lights was that they arrived just before dawn. This implies that the flight period is restricted to a narrow and perhaps dim time of day. Both males and females arrived at the lights, which is also unusual. In most sphingid species the males come to lights because they are on the prowl for females. We assume that both sexes came to the lights in their search for adult food plants, since that is a resource they both need (Janzen 1984a).

It is easy to determine the sex of an *O. kadeni* moth. Both sexes have a robust body with forewings that are only slightly longer and wider in the females. But the antennae of the male are thicker than are those of the female, and most diagnostic, they swell noticeably in the terminal third, making the male antennae markedly clubbed.

While you read this passage, a monster-eyed female moth of *O. kadeni,* about the size of a mouse, was scanning the ACG rain forest looking for an understory rubiaceous treelet with huge leaves, *C. macrocarpa,* on which to oviposit a single, giant egg. As soon as that egg is laid, it is in the sights of a parasitoid wasp hardly larger than a fleck of dust. The wasp is a member of the family Encyrtidae, which contains thousands of species, some of which attack Lepidoptera by placing their egg inside their host's egg. The encyrtid egg does not hatch until after the caterpillar has eaten and molted its way into its last instar. Then dozens to thousands of individual parasitoid larvae feed on the internal tissues of the caterpillar, eventually turning it into a mere shell of its former self, a vessel now full of wasp pupae.

ADULT VOUCHER: 04-SRNP-55887; JCM
CATERPILLAR VOUCHER: 04-SRNP-55971; JCM

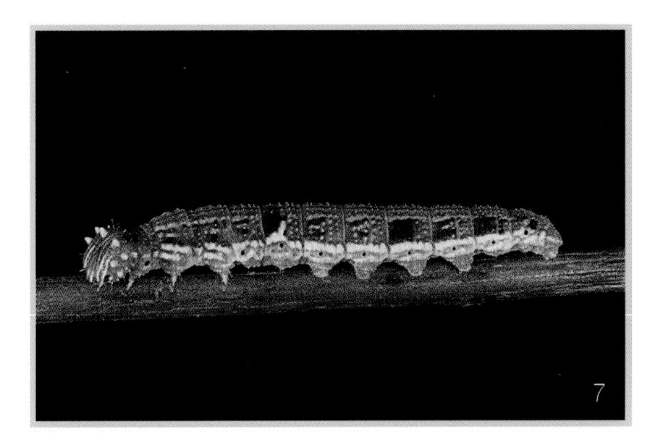
7

7. *FOUNTAINEA EURYPYLE* – NYMPHALIDAE

The ACG dry forest has its brown "*Memphis*"—*Anaea aidea* (#26)—and the ACG rain forest has its brown "*Memphis*"—*Fountainea eurypyle.* Why *Memphis* in quotation marks? Because were a taxonomist to ignore color and give serious weight to similarities in caterpillars, pupae, and food plants, it seems likely that both of these butterflies would be put in the genus *Memphis,* the species-rich, largely black charaxines with some iridescent blue on the upper side of the wings and bark-like patterns on the underside of the wings. Alternatively, perhaps *Memphis* should be broken up into many generic units. Then *Fountainea* and *Anaea* would seem to be justifiable taxa. Even then, the question would remain as to why there is just one of these brown (upper side) butterflies in each major ecosystem. There is a rare look-alike to *F. eurypyle* that occurs at apparently very low density in both ecosystems. The look-alike species, *Memphis chrysophana,* has a purplish sheen to the wings.

Caterpillars of *A. aidea* and *F. eurypyle* feed on several species of *Croton* (Euphorbiaceae). Again, one wonders if it is a coincidence that they both have light green caterpillars with very similar markings, and as such are easily distinguished from the variety of color patterns displayed by *Memphis* caterpillars. Although *F. eurypyle* is unambiguously a rain-forest butterfly, we do have four caterpillar records from deep in ACG dry forest, providing yet one more example of the widespread phenomenon of ACG rain-forest butterflies and moths that apparently circulate into ACG dry forest. Usual census methods for butterflies and moths do not pick up these occasional maverick adults, but if they leave eggs and those in turn produce caterpillars, then the *gusaneros* will find them.

ADULT VOUCHER: 02-SRNP-3518; JCM
CATERPILLAR VOUCHER: 05-SRNP-23744; JCM

8. *SIDERONE MARTHESIA* – NYMPHALIDAE

Staring idly into the top of a dry-forest oak tree on a long, hot afternoon in the dry season, one comes to the sudden realization that there is a brilliant red, flapping dot that appears periodically against the blue sky. For an hour it fools you into thinking that it is an *Agrias amydon* repeatedly launching and perching on the sap flow from a wounded *Quercus oleoides* branch. This branch seems festooned with what appear to be flaps of bark but are actually slightly tipsy *Memphis, Anaea, Archaeoprepona, Prepona, Historis,* and *Myscelis,* all butterflies congregated there to suck up the yeast-ridden fermenting sap (a.k.a., beer). In the process of finding other sources of fermenting sap, they transmit that same yeast to other newly damaged tree branches. But for some unknown reason, perhaps because it is shoved to the side by more aggressive butterflies, the red-winged butterfly keeps leaping into the air for several turns before re-alighting. Once a pair of binoculars is employed to telescope in on the red dot, it is pronounced to be *Siderone marthesia.* The underside, like that of the other sap feeders, resembles a dead leaf or a dead flap of tree bark, but the top bears a brilliant, eye-catching red (compared to blue for others, or even red and blue for *Agrias amydon*). The bright color says, once the bearer is in the air, "Don't bother to try. Yes, here I am, but I am way too fast." As with all the species that catch the human eye so readily, we really do not know to what degree the evolution of these colors as antipursuit signals has been intertwined with the evolution of courtship displays. Both functions are vital and seem to reinforce the expression of the bright color.

Although adult *S. marthesia* are butterflies of the canopy, only rarely descending to a rotten fruit or a fermenting, fruit-baited butterfly trap, the caterpillars are denizens of low, shrubby treelets in the forest understory—*Casearia sylvestris* primarily but also *Casearia nitida, C. tremula, C. arguta,* and *Zuelania guidonia.* All have long been treated as members of the tropical family Flacourtiaceae, but recent taxonomic changes have now placed them in the Salicaceae, the willow family, thereby giving Salicaceae a decidedly cosmopolitan distribution. Young caterpillars perch at the tips of leaves and look like a bit of trash hanging on the midrib of the chewed-upon tip. As they grow, they come to look more and more like a piece of rolled, dead leaf stuck by a bit of spider silk or fungal hyphae to a twig or green leaf. The bright green pupa hangs much like a small, green fruit among the foliage, and some two weeks later the brilliant red of the about-to-eclose butterfly is visible for twenty-four hours through the clear cuticle of the pupal wall. After hardening its wings for a few hours, the newly eclosed butterfly zips up to the canopy to begin its adult life, which may last six months or even more. The adults are present, but not reproducing, throughout the long dry season. They do not migrate away from ACG dry forest, though they do seem to be most abundant in or beneath the more prominent evergreen tree crowns in the forest along dry creek beds. Not just a dry-forest butterfly, *S. marthesia* also occurs in ACG rain forest but essentially nothing is known of its biology in that habitat.

ADULT VOUCHER: 04-SRNP-1457; JCM
CATERPILLAR VOUCHER: 81-SRNP-150; DHJ

9. *ASTRAPTES* INGCUP – HESPERIIDAE

As described for *Astraptes* LOHAMP in our portraits of ACG caterpillars (Miller et al. 2006), what had been known as *Astraptes fulgerator,* and still is by formal taxonomic rules, turned out to be ten different species when the DNA barcodes were compared with the variety of caterpillar color patterns, food plants, local distributions, and slight differences in adult facies (Hebert et al. 2004). One of the traits that really caught the ecologist's attention in *A. fulgerator* was that the caterpillars fed on the foliage of a great array of species in many dramatically different families and species. However, matching DNA barcodes with the food-plant records within and among caterpillars exhibiting certain similar morphological traits, the ten species within the *fulgerator* complex turned out to be as highly host specific as are all the other hundreds of reared species of ACG pyrgine hesperiid caterpillars. In particular, *Astraptes* INGCUP (our temporary and unofficial name), displayed a rare trait that appears here and there in the inventory project: It has just two groups of food plants that are taxonomically very different. From the viewpoint of *A.* INGCUP caterpillars and ovipositing females, three species of ACG *Cupania* (Sapindaceae) and ten species of ACG *Inga* (Fabaceae) are food plants that do have something in common; but to the botanist or plant chemist, they have nothing in common. To date, no other species of ACG specialist caterpillars have been found to share just these two food plants. The restricted diet of *A.* INGCUP does not translate into an environmentally limited distribution. The species occurs in dry forest, rain forest, and their intergrades.

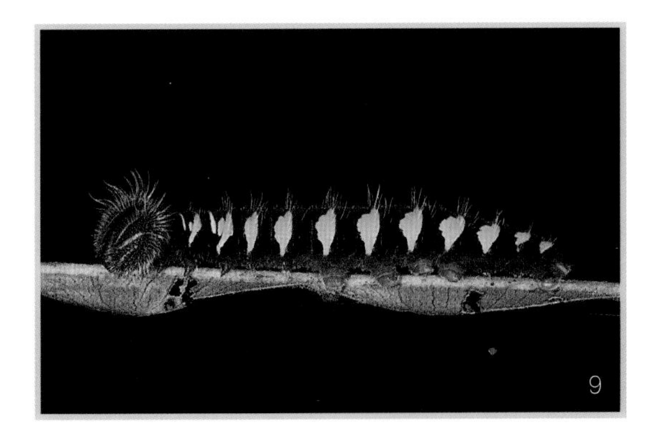

Genetic variation, adaptation, disruption of gene flow, and speciation go hand in hand, and the *A. fulgerator* story is an excellent example of that. Furthermore, *A.* INGCUP illustrates what may be a very early phase in the process of speciation. After barcoding eighty-three adults of reared *A.* INGCUP, one-third of which were caught in the wild feeding on *Cupania* and the remainder on *Inga,* we can say with certainty that there is no DNA barcode difference between specimens from the different food-plant families. The caterpillars, yellow to deep orange, strongly ringed, with white hairs and black faces, do show one kind of variation. The rings are sometimes broken on the back, sometimes even making vertically elongated side spots. But the broken-ring individuals are found on both food-plant families, and in both dry forest and rain forest. Nothing in adult facies suggests that *A.* INGCUP is in fact two species. However, in view of these two very different food plants, there still remains the possibility that there are two cryptic species hiding in one DNA barcode, and only further study will resolve the matter.

The strikingly marked black, white, and blue iridescent color pattern of *A.* INGCUP is part of a mimicry complex containing many tens of ACG species and easily five times that many in the Neotropics. As with the black, white, and blue pattern of *Jemadia/Phocides/Elbella* (see *Phocides nigrescens,* #91 and *Parelbella macleannani,* #92), it probably sends the message, "Don't bother to try," to potentially foraging birds. The widespread nature of this mimetic complex, with many lineages having evolutionarily joined over the millennia, suggests that it functions quite well. It certainly forms a major part of the answer to the question of what has kept the ten species within *A. fulgerator* looking so much alike while their caterpillar color patterns and food plant uses have diverged.

ADULT VOUCHER: 93-SRNP-3048; JCM
CATERPILLAR VOUCHER: 05-SRNP-19576; JCM

10. *ASTRAPTES* YESENN – HESPERIIDAE

Astraptes YESENN is another member of the ten-species complex that was detected and split out of the *Astraptes fulgerator* species complex by a combination of comparative morphology, caterpillar natural history, and DNA barcoding (Hebert et al. 2004). As can be seen by comparing the male of *Astraptes* YESENN with *Astraptes* INGCUP (#9), the adults are essentially identical, though with experience some of the ten species can be separated from others by their adult facies. But the caterpillar of *A.* YESENN both looks quite different from that of *A.* INGCUP and feeds on quite different plants. Whereas *A.* INGCUP feeds on foliage of *Cupania* and *Inga* (Sapindaceae and Fabaceae), *A.* YESENN is a specialist on the foliage of *Senna* and a few other Fabaceae. In the ACG, its food plant is usually the rain-forest *Senna papillosa,* but in the intergrade between the dry and rain forest it also feeds on *Senna hayesiana,* which appears to be the dry-forest analogue to *S. papillosa.*

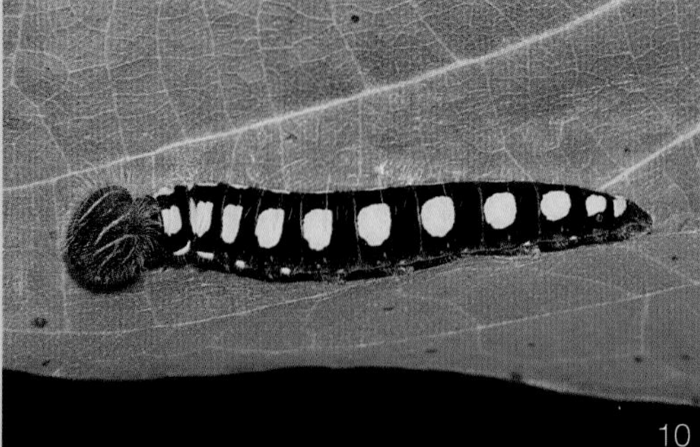

The distinctive caterpillar of *A.* YESENN, black with yellow side spots, appears to have evolved just once in this complex, but the look is borne by three species: *A.* YESENN, *A.* LOHAMP (Miller et al. 2006), and *A.* LONCHO. The caterpillars of these three species are essentially identical. Also, they are sympatric to parapatric: *A.* YESENN and *A.* LOHAMP occupy all of the intermediate-elevation rain forest in the ACG down to its lowest extent (elevation of 200 meters), whereas *A.* LONCHO feeds on the upper-elevation *Lonchocarpus* of Volcán Cacao (and probably the other ACG volcanoes as well); *A.* YESENN fully overlaps with *A.* LOHAMP, which follows its food plant *Hampea appendiculata* right up into the lower cloud forest, while *A.* YESENN feeds on *Senna hayesiana* and *S. papillosa* as these plants follow human disturbance up the sides of the volcanoes. All three of these black caterpillars with yellow side spots can be found in the same hectare at Estación Cacao (elevation of 1,100 meters) on the upper western side of Volcán Cacao.

How do the adults of these three species of *Astraptes* tell each other apart? First, perhaps they do not do it very well. That might generate hybrids, but the DNA barcoding does not yet suggest this to be the case. Such hybrids would DNA barcode as the mother but, at least theoretically, they might be found feeding on the food plants of both parents. Only further observation and experimentation will tell. It is not known if the three species of caterpillars could or would feed on each others' food plants if the eggs were placed there. Alternatively, they may discriminate among themselves very well, perhaps by differences in courtship behavior or by morphological differences not apparent to us, such as discriminating between differently reflected wavelengths of light. In full sunlight, the blue of the upper side of the wings of *A.* LONCHO (which appears essentially identical to *A.* YESENN and *A.* INGCUP) is deeper and more sapphire than the bluish-turquoise of the other two species. That should be quite enough for *A.* LONCHO to be identified as such by the other two species, and vice versa. There is much more to learn about this intriguing complex of very similar species.

ADULT VOUCHER: 99-SRNP-11; JCM
CATERPILLAR VOUCHER: 04-SRNP-41644; JCM

11. *CHRYSOPLECTRUM PERVIVAX* – HESPERIIDAE

Chrysoplectrum pervivax is one of those butterflies that you just know is there but never see, and not because it is easily confused with the similar adults of various species of ACG *Astraptes* (see *A.* INGCUP, #9 and *A.* YESENN, #10). Perhaps it is not seen because it is nocturnal or crepuscular? Not likely—it lacks the red eyes so common in many crepuscular Hesperiidae, and it has never been seen at light traps. As a dry-forest butterfly, we do not expect it to live only in the tree canopies high overhead, but that might be the cause of the apparent absence of adults. Another possibility is that owing to its very tight restriction to a single species of rare and local tree for its caterpillars' food plant, it stays only in the vicinity of those plants; therefore, unless you look there, you will never find it. Whatever the case, this purely dry-forest butterfly is notable for remaining unseen, as are many species of ACG rain-forest Hesperiidae.

So how do you know it exists? It leaves its tracks in the form of caterpillars feeding on their sole ACG food plant, *Triplaris melaenodendron* (Polygonaceae). This species and several equally rare species of *Coccoloba* are the only ACG dry-forest trees in this plant family. The large-leafed *T. melaenodendron* is a member of the riparian vegetation in ACG dry forest. Its wind-dispersed seeds are surely carried far away from riparian vegetation, but the species seems to survive to mature status only along watercourses. This may not be due to a simple need for roots to be in moist soil. Another special feature of *T. melaenodendron* is that it is an ant plant with hollow stems that are fiercely (and apparently obligatorily) patrolled by a single species of *Pseudomyrmex* ant, the same genus that occupies the ant acacias (Janzen 1967). How *C. pervivax* caterpillars survive

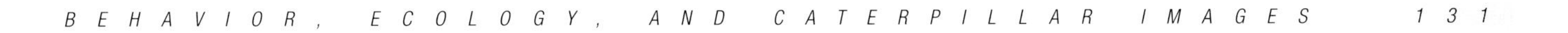

the patrolling of these ants, if they do survive them, is not at all clear. The most common evidence of the presence of female *C. pervivax* is the small houses comprised of a leaf folded over made by the first and second instars and, much more rarely, by the later instars with their brilliant black, white, and orange color pattern.

The ostentatious color pattern of the caterpillar is presumably part of the same mimicry process ascribed to other ACG caterpillars with those color patterns, but there is more. In the same ACG habitat there is a hesperiid, *Polythrix caunus,* that is similar as an adult to *Urbanus belli* (brown with long tails, #77), but not as a caterpillar. The adult *P. caunus* bears no resemblance whatsoever to the adult of *C. pervivax,* but the caterpillars are extremely similar, as are the pupae. The caterpillar of *P. caunus* feeds on the mature foliage of bignoniaceous vines. This raises the possibility that these two adults who look so different are closely related species. If validated, such a relationship reinforces the concept that each adult morph, the black with iridescent blue and the brown with long tails, is a representative of a very different mimicry complex, in spite of a recent common ancestor. Yet the common ancestry is still revealed through the morphology of the caterpillars and pupae.

ADULT VOUCHER: 03-SRNP-599; JCM
CATERPILLAR VOUCHER: 92-SRNP-183; DHJ

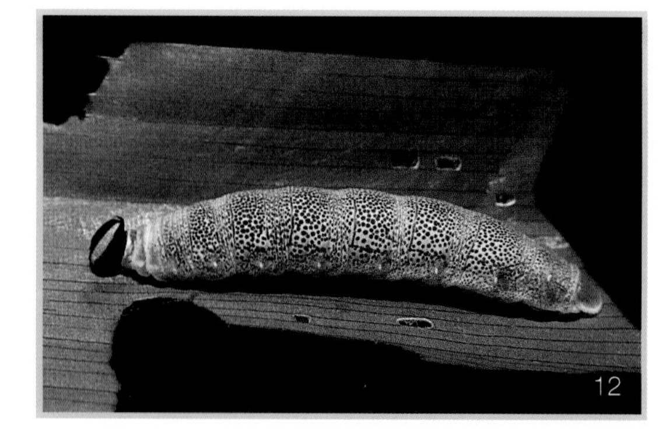

12. *NEOXENIADES MOLION* – HESPERIIDAE

Compare *Neoxeniades molion* with *Astraptes* INGCUP (#9), *Astraptes* YESENN (#10), and *Chrysoplectrum pervivax* (#11). This comparison includes only four species, a small portion of the mimicry complex discussed under *A.* INGCUP. The complex consists of species from various subfamilies: *N. molion* is a hesperiine Hesperiidae, its caterpillars and those of its ancestors feeding on monocots—grasses, palms, bananas, gingers, and so on; *A.* INGCUP and *C. pervivax* are pyrgine Hesperiidae, their caterpillars and those of their ancestors feeding on dicots—the broad-leafed woody plants that make up most of the forest biomass. The hesperiine line separated from the pyrgine line many tens of millions of years ago. Even back then there could have been a color pattern such as this, but it is a total stretch of reason to suggest that these four butterflies, and many tens of other unrelated Neotropical species, look similar because they have a common ancestor and have maintained the color pattern since. We should add that such a line of reasoning would suggest a link between this particular pattern and all other color patterns in Hesperiidae. Such a connection is not likely. Imagine for a moment that you are a bird trying to tell these species apart as they are buzzing in flight many meters away, darting in and out of the vegetation. Next, concede for the sake of argument that insectivorous birds cannot tell them apart any better than you can. That is exactly how mimicry works.

But this view of having a color pattern in common does bring to mind the question of how the adult butterflies distinguish among themselves, at least with respect to courtship and territoriality? First, it is likely that the combination of very slight differences in the shade and intensity of blue, quality and exact pattern of white, and shade and intensity of chocolate to black all interact to create a species-specific image. First, the male or female of a given species has only to recognize its own kind, rejecting all others. Second, in addition to the visual spectacle of a three-dimensional black, white, and blue fuzzy object, it is reasonable to expect that these butterflies, like probably nearly all other Lepidoptera, emit pheromones when in close range, odors that help one species identify its own kind. Third, the magic of matching genitalia at the right time with the right behavior probably serves as an effective barrier against inter-specific mating, should it get to that point (which it probably rarely does). These species may look alike to humans, but the morphology of the genitalia of *N. molion* is quite different than that of *A.* INGCUP, *A.* YESENN, or *C. pervivax.*

ADULT VOUCHER: 00-SRNP-10565; JCM
CATERPILLAR VOUCHER: 98-SRNP-3076; DHJ

13. *EUEIDES PROCULA* – NYMPHALIDAE

In the 1980 rainy season, only two years into the caterpillar inventory project, the twenty- to thirty-meter tall *Ehrblichia odorata* (Turneraceae) trees in the ACG Bosque Humedo, a semi-evergreen patch of old-growth forest in Sector Santa Rosa, were abruptly and totally defoliated by *Eueides procula* caterpillars feeding *en masse* on these trees, their sole food plant (Janzen 1983). This was a surprise not only because the defoliation was by the caterpillar of a butterfly, a rare event in contrast to defoliation by moth caterpillars, which is a common event, but also because this group of nymphalids is normally thought of as feeding on herbs and vines. But then again, Turneraceae is generally thought of as a family of herbs. The leafless *E. odorata* crowns stood out dramatically against the blue sky, nestled among the fully leafed crowns of the other tree species. The defoliation seemed to have started in the crowns of the biggest trees and then worked its way to the saplings and even seedlings in the understory as the starving, restless caterpillars fell or dropped off and wandered in search of food. White pupae were silked to leaves everywhere. Soon thereafter the air seemed choked with adult *E. procula* flying, chasing, courting, mating, and setting the stage for the cycle to repeat itself.

13

In the twenty-seven years following this two-generation population explosion, *E. procula* has been an occasional but-

terfly in the ACG zone of interdigitation of dry and rain forest. Never common, one would not suspect that it could be a major defoliator. What led to the 1980 population outbreak will never be known, though the 1979 collection records showed higher densities of immature *E. procula* than any year after 1980. It seems clear what knocked the population back: There were quite literally no *E. odorata* leaves remaining to serve as oviposition sites, and surely the great bulk of the resulting butterflies went to their graves leaving no offspring. Why the population has never again grown to such high numbers on this common ACG tree remains a total mystery. In general terms it is a common event for a species to become extremely abundant one year and then remain at a low and inconspicuous density for decades in the ACG dry forest (Janzen 1981).

ADULT VOUCHER: 03-SRNP-18273; JCM
CATERPILLAR VOUCHER: 80-SRNP-283; DHJ

14. *TITHOREA TARRICINA* – NYMPHALIDAE

At first glance *Tithorea tarricina* is just one more of the many species that are toxic-to-eat Mullerian mimics, flapping and sailing along a rain-forest trail, clearly showing their rusty orange-red hindwings and black forewings with yellow spots. A comparison of *T. tarricina* with *Eueides procula* (#13) illustrates how similar two noncongeneric species can be. When including other species in the comparison, it quickly becomes obvious that this pattern is employed by dozens of species of Mullerian mimic ithomiine and heliconiine nymphalids, pericopine arctiids, and probably others, with scattered Batesian mimics not only in Costa Rica but throughout the Neotropics. For this comparison, browse the plates of *Butterflies of Costa Rica* (DeVries 1987) or William Haber's Web site for the butterflies of Monteverde (found within http://www.cs.umb.edu). As has been said before and will be said again, in addition to the classical concept of mimicry as a phenomenon based largely on learning, it may well be that the birds are even genetically hard-wired to ignore this color pattern (Smith 1975, Janzen and Pond 1976).

The story can be expanded into North America. In 2005, an entire issue of the journal *Nearctic Lepidoptera* was devoted to the regal fritillary, *Speyeria idalia* (Ross 2005). This very large and attractive endangered species in the United States once ranged widely in the northern half of the country east of the Rocky Mountains (see www.butterfliesandmoths.org for further information). It occurred just about everywhere violets, *Viola* (Violaceae), the caterpillar's food plant, were found in the tallgrass prairie. Now its distribution is reduced to a few enclaves dotted over that once large area. For those familiar with the U.S. butterfly fauna, *S. idalia* is an outstanding color pattern, not duplicated by other species in Kansas, Iowa, Indiana, or the Dakotas. Mimicry may never cross your mind, yet *S. idalia*'s coloration may remind you of something. The butterfly has relatively slow, flapping, and ostentatious flight. In flight, it is a blur of orange and black with white, off-yellow, and silver spots. When perched on flowers with its wings wide open, advertising itself to the world, the dorsal surface of the hindwings display black with white spots while the forewings display orange with some scattered black etchings. In short, if you reverse the fore- and hindwing colors, *S. idalia* resembles *T. tarricina* or *E. procula.* But you say that the predators of the *T. tarricina* Mullerian mimicry ring are in the Neotropics, not the plains of North America. But keep in mind that in the spring, when the birds migrate north to the Midwestern United States, they leave behind neither their genetically based bird instincts nor their learned experiences.

ADULT VOUCHER: 05-SRNP-3553; JCM
CATERPILLAR VOUCHER: 00-SRNP-1248; DHJ

15. *XYLOPHANES TYNDARUS* – SPHINGIDAE

There are amazingly few species of truly green moths and even fewer green species among the butterflies. They do, however, provide a stunning image, and therefore their portraits are shown disproportionately more often than the number of green species would warrant. We certainly favor them in this book (see *Moresa valkeri,* #1; *Oryba achemenides,* #5; and *Oryba kadeni,* #6). It seems odd that green adults are in a minority, since a majority of the forested world is green. It may be that a green moth does not look much like a green leaf or leaf part. Even though there seem to be so very many kinds of leaves, each is unmistakably a leaf in shape, luster, margin, position, size, and other more subjective traits. In other words, a green something that is not a leaf actually does not look much like a leaf. *Xylophanes tyndarus* is one of the exceptions. The species is rare at light traps throughout its range and rare as caterpillars unless you extensively search their specific rubiaceous *Faramea* food plants. When resting in the daytime, it hangs by one leg from under a leaf or perches multilegged on a green stem, and appears to be a somewhat oddly shaped leaf. We know of no experiments evaluating the cryptic nature of the adult, but it is a fair guess that it would actually fool many species of birds that search the foliage visually.

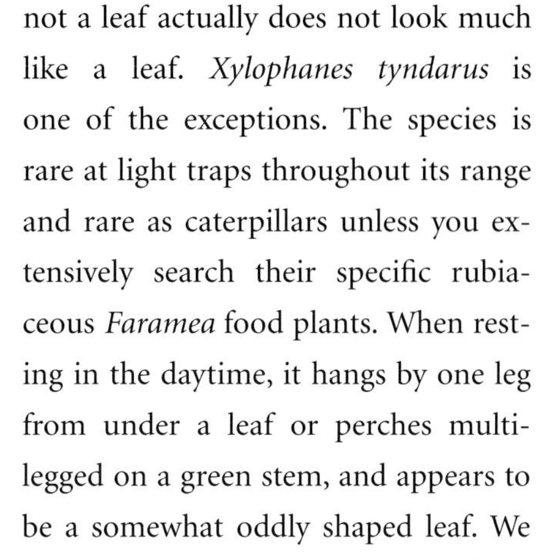
15

Although *X. tyndarus* is a proper *Xylophanes,* that is, its caterpillar feeds on Rubiaceae as expected of species in this genus of moths, it appears to ignore the multitudes of species of *Psychotria* used as food plants by so many other species of *Xylophanes.* Its 216 rearing records by the caterpillar inventory project are all from *Faramea occidentalis,* a distinctive understory treelet that occupies the interface between ACG dry forest and ACG rain forest. It shares this food plant with *X. chiron* (Miller et al. 2006) and *X. pistacina,* but these two species are not specialists and utilize a variety of other rubiaceous food plants.

ADULT VOUCHER: 03-SRNP-2847; JCM
CATERPILLAR VOUCHER: 98-SRNP-4995; DHJ

16. *ADHEMARIUS YPSILON* – SPHINGIDAE

Nine years ago we took a large number of ACG staff on a nature-watching weekend hike in the forest. We looked at things in Santa Rosa, including the rearing barn, and passed around some caterpillars. Two days later someone in the group pointed toward a large *Adhemarius ypsilon* caterpillar eating the top out of a two-meter-tall lauraceous sapling in the dark understory of tall forest. At this particular moment, we were a group of some thirty staff and their children—talking, looking, and communing with nature. We asked the crowd if anyone knew what it was, pointing in the direction of the highly patterned and highly cryptic caterpillar. We fully expected a simple reply, such as "*gusano*" (caterpillar). Abruptly, from easily five meters away, we heard, "Creo que es *Adhemarius*" ("I think that it is *Adhemarius*"). A teenage boy, nervous over the sound of his own voice, had called out over the murmur of the crowd. A new *gusanero* was found. Harry Ramirez, the son of an ACG truck driver and former park guard, had seen this caterpillar in a plastic bag among hundreds of others just once two days earlier, at which time we must have offhandedly called it by its scientific name. Now, over ten thousand caterpillar rearings later, Harry is well into his lifelong career as a *gusanero* (Janzen 2004), a job he accepted after leaving a meat-packing plant and taking a 30 percent cut in salary because his father told him that as a *gusanero,* "you will learn things." He is still identifying caterpillars from five meters away.

16

Adults of *A. ypsilon* are among the most regular sphingid arrivals at a light trap in ACG rain forest at any time of the year. The adult is larger than its congeners: *A. fulvescens* in the cloud forest, *A. daphne* in the lowlands, and *A. gannascus* (the name that used to be used for all three) in the rain forest. The adult of *A. ypsilon* has a species-specific pattern on both the front and hindwings. This pattern, when the moth is at rest, resembles a torn, rotten leaf, a nice appearance while sedentary. When disturbed at warmer daytime temperatures, the moth will launch into flight with the red-pink hindwings serving as an attention-grabbing signal, a signal that abruptly disappears among the real torn and rotten leaves when the moth alights. Many sphingids and noctuids use this escape mechanism, which brightens up many a moth portrait, including some of the species in this book (*Gonodonta pyrgo,* #74; *Azeta rhodogaster,* #89; *Ferenta castula,* #45). But for anyone looking casually at *Adhemarius* in Costa Rica, we caution that there is a rarely seen look-alike with slightly more falcate wing tips, *Orecta venedictoffae,* which can be seen on the ACG inventory project Web site (go to http://janzen.sas.upenn.edu). The only evidence of the presence of *O. venedictoffae* in Costa Rica, a very unexpected species known previously only from Ecuador, is a photograph taken by James Sogaard at Estación Pitilla in the ACG rain forest.

ADULT VOUCHER: 02-SRNP-1488; JCM
CATERPILLAR VOUCHER: 04-SRNP-3129; JCM

17. *PARIDES IPHIDAMAS* – PAPILIONIDAE

Look at the butterfly on the page facing adult #17. Now imagine that these two butterflies are buzzing side by side down a forest trail, stopping at a flower here and there. No bird in the world will be able to discriminate between *Parides iphidamas* and *Mimoides euryleon.* The wings are velvet black, each festooned with a red and a white patch, markings that telegraph, "Look at me! I will give you a monster bellyache if you should ever make the mistake of eating me." At any one place in the ACG, from three to ten species belonging to this huge Mullerian and Batesian mimicry complex can be encountered flying at the same time. However, the number of species visiting the same flowers, ovipositing on the same food plants, puddling on the same mud, or coursing at the same height off the ground is usually reduced to one to four species. In other words, the observer can use more than the color pattern to determine what species is in sight. But even then, there are many times when the identity is known for certain only if the butterfly is captured on film or by hand. For example, look at *P. iphidamas,* and take note of the white margins to the tiny bays between the bumps on the posterior margin of the hindwing. The female of the fully sympatric *Parides arcas* is essentially identical, except that the margins are pink instead of white.

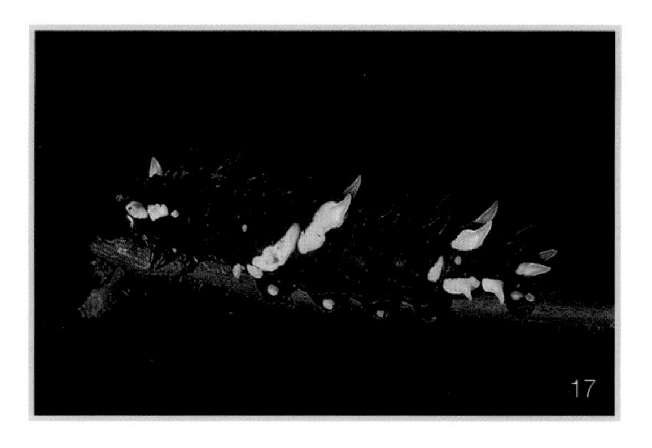

There are at least five species of *Parides* with black-white-red females (and sometimes males) in the ACG lowlands. Because of their bright and continually ostentatious colors, and because their caterpillars feed only on the foliage of *Aristolochia,* vines in the Aristolochiaceae that are famous for being rich in the toxin aristolochic acid, *Parides* are viewed (probably correctly) as the models for the swirl of black-white-red Batesian and Mullerian mimics (see *Mimoides euryleon,* #18) found throughout the lowland Neotropics. Many other species feeding on Aristolochiaceae are also brightly colored and are inedible models for a mimicry complex (e.g., the pipevine swallowtail of the southeastern United States). However, the entire discussion of mimicry centered around *Parides* or any other set of models should always take into account the predators that do not have to learn anew every generation that the ostentatious model is inedible but instead are genetically programmed to avoid these color patterns. Susan Smith demonstrated this phenomenon with the bright yellow, red, and black rings of coral snakes and their potential avian predators (Smith 1975). Another caveat that is not comparatively studied is that a given "inedible" or "toxic" insect may be lethal or vomit-inducing to one species of potential predator, but a great meal for another species.

The ostentatious caterpillars of *Parides,* with black to deep maroon and white patches or stripes, and other *Aristolochia*-eaters in the ACG are clearly in some other caterpillar mimicry complex of their own, and the above caveats and considerations apply to them as well. It is widely believed that the toxins that protect the adults from being hawked out of the air by agile flycatchers and jacamars are acquired from the food plant and sequestered by the caterpillar, and this is probably true. But when it comes time to pupate, they turn into cryptic, yellowing green leaves, somewhat wilted, silked to the side of a bare twig and quite invisible against the foliage. It is striking that none of these species has gone the route of aposematic coloration in the chrysalis.

ADULT VOUCHER: 89-SRNP-498; JCM
CATERPILLAR VOUCHER: 04-SRNP-2229; DHJ

18. *MIMOIDES EURYLEON* – PAPILIONIDAE

The butterfly we currently recognize as *Mimoides euryleon* was previously placed in other genera. We learned this butterfly first as *Papilio euryleon.* Then it became known as *Eurytides euryleon.* Finally it was placed in *Mimoides* so that the genus would be monophyletic—all members of the group descended from a common ancestor, and all those descended from that ancestor placed in the same genus. Why bother with this phylogenetic splitting of hairs? In general terms, the more reliably monophyletic the group, the more accurate the predictions about the unknown biology, including morphology, of other members of the group. For example, the distinctive fat, green pupae of species of *Mimoides* are similar, as are the patterns in the wing veins and the DNA barcodes. As species adapt to a variety of environmental conditions, they tend to diverge in certain traits. For example, the Annonaceae-eating caterpillars of *M. euryleon* occupy upper ACG elevations, whereas caterpillars of *M. branchus* occupy lower ACG elevations, and they do not share the same color pattern or morphology.

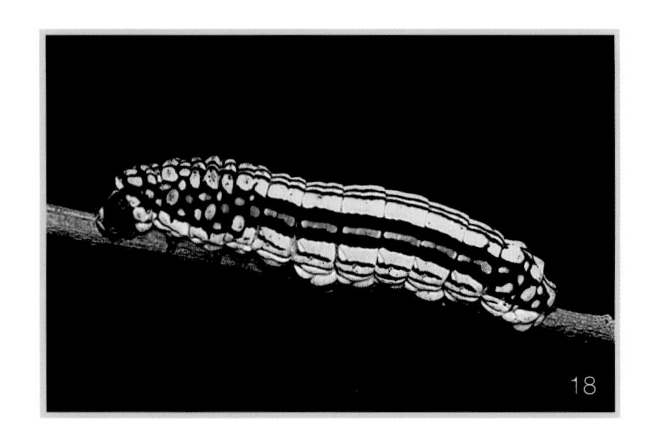
18

It is unlikely you could walk for an hour in the rainy season along an ACG trail in insolated, or sunlight-exposed, secondary succession without encountering a large, black butterfly with white patches on the forewings and red patches on the hindwings. It has a sufficiently fast wingbeat such that the details of the white and red are blurry, but the colors are obvious. The butterfly pauses here and there on foliage, tasting for the chemicals that suggest a potential site to lay an egg. It visits butterfly flowers, like the blue flowers of *Stachytarpheta frantzi* (Verbenaceae), that are high on the list of favorite nectars. But what species are you observing? There are at least fifteen species of ACG mimetic Papilionidae with this behavior and color pattern (compare with *Parides iphidamas,* #17).

Many of these mimics are so similar that even those who devote their lives to the study of butterflies must catch the specimen to ascertain its correct identity. Not all fifteen species visit the same flowers at the same time in the same ACG ecosystem, but since the birds circulate, this may not matter much. Intermingled with them are look-alikes from other families, such as day-flying pericopid moths, *Dysschema jansonis,* Arctiidae; pierids, *Archoneas tereas;* and a host of nymphalids with red to orange on the hindwings and white to yellow on the forewings. The nymphalids do not match the textbook examples of exact mimetic resemblances of the papilionids, pericopids, and pierids, and are generally ascribed to other mimetic complexes. But when in flight in bad light and to a not very mentally or visually discriminating bird, they certainly can easily be ascribed to the margins of the *Mimoides* mimetic color regime. The principal model of this Mullerian and Batesian mimicry complex seems to be *Parides.*

ADULT VOUCHER: 00-SRNP-9501; JCM
CATERPILLAR VOUCHER: 04-SRNP-22947; JCM

19. *BARDAXIMA PERSES* – NOTODONTIDAE

Bardaxima perses is a very annoying moth, if not challenging. Although it is a very rare animal in ACG light traps, its brilliantly colored caterpillar can be reliably found in the rainy season eating its sole dry-forest food plant, *Ouratea lucens,* the only dry-forest species of Ochnaceae. We assumed this moth followed a classical pattern of being a highly host-specific and well-defined species of ACG notodontid. We based this assumption on the dietary habits of two distinctive species of Hesperiidae, *Udranomia kikkawai* and *U. orcinus,* which also utilize *O. lucens* as their dry-forest food plant.

New knowledge changes old impressions. As the caterpillar inventory project expanded into the rain forest in the late 1990s, we began to search the huge leaves of the tree *Cespedesia spathulata,* another Ochnaceae with little vegetative resemblance to the dry-forest understory shrub *O. lucens* other than the similarity of their yellow flowers. While somewhat of a surprise, it made phytochemical sense to discover that *U. kikkawai* and *U. orcinus,* which we had come to think of as dry-forest butterflies, were also in the rain forest, dutifully feeding on new *C. spathulata* leaves, just as they eat the very new leaves of *O. lucens* in dry forest. Then we found the brilliant and distinctive caterpillars of *B. perses* feeding on the same *C. spathulata* leaves. A nice phytochemical story was beginning to unfold. Soon thereafter the caterpillars of both species of *Udranomia* were found feeding on another odd rain-forest species, *Quiinia schipii* (Quiiniaceae). The Ochnaceae and Quiiniaceae are not related taxonomically in any obvious way. It is a safe bet that we will find their foliage has some secondary defensive compound in common. The pattern in food-plant relationships emerged further when *B. perses* caterpillars were found feeding on *Q. schipii* as well. This observation was followed by the discovery of *B. perses* caterpillars on the foliage of a second species of Quiiniaceae, *Lacunaria panamensis.* Now, of course, the prediction is that eventually caterpillars of *Udranomia* will be found on the foliage of *L. panamensis.*

20

Their dietary patterns were revealed through expanded studies, and eventually, we were able to use DNA barcoding to assist us in identifying ACG Lepidoptera. Our first targets in the search for cryptic species were the cases in which a well-known, morphologically defined species occurred in both ACG rain forest and dry forest, but showed no morphological differences in specimens from these two adjacent ecosystems. *Bardaxima perses* fit the bill. A preliminary survey, which now needs to be repeated in substantial depth, showed that what we were calling *B. perses* is actually two sympatric species, each occurring in both rain forest and dry forest, at least according to nucleotide sequences of the COI gene in the mitochondrial DNA. This means not only that at least one of the two is probably an undescribed species, but also that at present we do not know which of the two, if either, matches the holotype specimen of *B. perses.* In fact, both species may be new because *B. perses* was described in 1900 (Druce 1900) based on a specimen from Manaus, in Amazonian Brazil, and it would not be surprising to discover that it is not the same species we have in the ACG. The discovery of cryptic species through DNA barcoding technology is also occurring in other groups: *Astraptes fulgerator* (Hesperiidae), *Xylophanes porcus* and *Xylophanes libya* (Sphingidae), and *Automeris zugana* and *Eacles imperialis* (Saturniidae).

ADULT VOUCHER: 03-SRNP-13555; JCM
CATERPILLAR VOUCHER: 03-SRNP-34559; DHJ

20. *CALLEDEMA PLUSIA* – NOTODONTIDAE

Your eye is immediately drawn to the silver-white stripe prominent on the forewing of *Calledema plusia,* a moth that looks ever so Notodontidae in its color patterns, wing shapes, and body-length to wing-length ratios. What does it mean to say it looks like a member of Notodontidae? Well, if you were to sort hundreds of thousands of specimens of mounted moths collected at ACG light traps, a notodontid *gestalt* would emerge. This particular gestalt would allow you, almost instinctively, to place a species you have never seen before in the correct family, in the example here, the Notodontidae. A different gestalt exists for noctuids, geometrids, limacodids, lasiocampids, saturniids, sphingids, and so on.

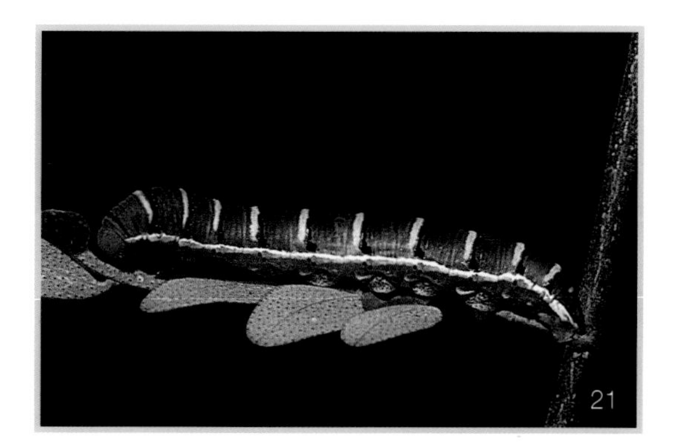

But why the silver-white stripe on the upper side of the forewing of *C. plusia?* In the museum drawer it is eye-catching and is the basis for its species name "*plusia*," making a comparative reference to the subfamily Plusiinae in the Noctuidae, many of which have metallic white, silver, or gold markings on their forewings. While the moth perches during the day in its normal habitat, the silver resembles a long, thin, slit-shaped hole through a dead leaf. In its museum-spread position, the moth is highly symmetrical, a pose that almost never occurs in nature. Its shape would be a dead giveaway to a foraging bird if the moth were to be seen as you see it in our portrait. The living moth sits with its forewings folded like a tent over its back, making it both three-dimensional and asymmetrical from the side. Since most birds possess monocular vision and probably see the world through one eye or the other, they are most likely to see this moth from one side of the tent or the other. If you close one eye and focus your attention on just one forewing, it is asymmetrical and resembles a torn leaf.

The caterpillar of *C. plusia*, which resembles a twisted, dead, dry leaf, is using the same ploy as the adult but in a very different world. When the caterpillar is feeding on the leaf blade, it perches exactly on the margin being cut away. As such, not only does its variegated, light-brown body match the pale color of most of the underside of the leaves of its Chrysobalanaceae food plants, but by its position it appears to be a ragged, torn, dry edge of the leaf. The cryptic nature of this morphology is further reinforced by behavior. When disturbed, the caterpillar lifts its head and tail end off the substrate, giving the appearance of a dangling, curled, dead leaf.

ADULT VOUCHER: 04-SRNP-3550; JCM
CATERPILLAR VOUCHER: 04-SRNP-15318; JCM

21. *ZERENE CESONIA* – PIERIDAE

Dan Janzen grew up on the outskirts of Minneapolis, before there were suburbs, in the early 1950s. He still remembers his first field-collected and reared caterpillar, a green worm feeding on the leaves of a shrubby legume in a glade in a city park. He nurtured this caterpillar until it developed into a chrysalis (a foreshadow of his later endeavors as a *gusanero*). Out of the chrysalis came a fine yellow southern dogface butterfly, *Zerene cesonia*. Obviously, Minneapolis is a long way from Costa Rica ecologically, geographically, or even as the crow flies. We know *Z. cesonia* is not migrating back and forth as do monarchs: It is a full-time resident of the southern United States. Its population creeps north each summer, only to be blasted back by the blizzards each winter. Scattered across the entire intervening landscape between Minnesota and Costa Rica, *Z. cesonia* is a resident breeding population. In Costa Rica it occurs in the dry environments, where shrubby legumes are common. Although it may follow pastures into the ACG rain forest on occasion, it needs human-associated disturbance to make the rain-forest ecosystem habitable.

The males of *Z. cesonia* all seem to be the color and the pattern shown here in the gallery. However, as is usual among species of Pieridae, the females range from pale to dark yellow and have a highly variable light to dark black pattern. The biological significance of the small eyespot on the dorsum and venter (underside) of the forewing is unclear. The spot has much in common with spots on the wings of many other medium-sized pierids. Seemingly, the function is shared among many species. The spot is probably not evoking a startle response among predators, as do the false eyespots on forest caterpillars and pupae, and even the wings of many species of ACG Nymphalidae.

ADULT VOUCHER: 02-SRNP-32314; JCM
CATERPILLAR VOUCHER: 02-SRNP-32305; DHJ

22

22. *EPIA MUSCOSA* – APATELODIDAE

This is an essay about taxonomic meandering. Biologists redefine and reshape taxonomic units constantly, and not just at the species level. For instance, *Epia muscosa* is a cocoon-spinning caterpillar with a nonfeeding adult, as are all Apatelodidae, which made it very easy to place it into the Bombycidae, where *Epia*, *Colla*, and *Quentalia* previously resided for a very long time. However, it has now been decided to restrict Bombycidae to the Old World tropics. This means that *Epia* becomes a member of the apatelodid subfamily Epiinae. If all this is confusing to you, it is even more confusing when keeping track of the natural history of these animals and relating them to others that have similar biology.

Epia muscosa is just a small rain-forest saturniid. It spins a strong silk cocoon, as do many Saturniidae and all Bombycidae; passes long periods as a dormant pupa within the cocoon, even in the rain forest; lives only a few days as a non-feeding adult (it does not have functional mouthparts); and has much larger and heavier females than males. The female is laden with her full egg load at the time of eclosion, mates during the first hours of the night of eclosion, and lays most of her eggs in the evening of the following night. In behaving this way, she displays the behavior of the other members of Apatelodidae, and Oxytenidae as well. The Oxytenidae (see *Oxytenis modestia* #47) are yet another saturniid-like and species-poor bombycoid family that was recently placed by some taxonomists as a subfamily in the Saturniidae.

Taxonomy is also meandering on the food-plant side of this story. It used to be that there was a family of plants, the Moraceae, into which were placed the figs (many *Ficus* spp.), cecropia (*Cecropia* spp.), *Coussapoa*, and a batch of other species. Then the *Cecropia* and *Coussapoa* were split out into the Cecropiaceae. So now we have the caterpillars of the mossy green and brown adult of *E. muscosa* eating one family, Cecropiaceae, and the caterpillars of the white and dark bird-dropping look-alike of *Colla rhodope* eating the other, Moraceae. However, the last instars of these two genera are nearly identical in form and color. Also, the earlier instars, obviously cryptic in resembling some kind of mold-ridden twig, have the same form, even if their color patterns are a bit different.

ADULT VOUCHER: 04-SRNP-2970; JCM
CATERPILLAR VOUCHER: 99-SRNP-13008; DHJ

23. *COSMOSOMA CINGULATUM* – ARCTIIDAE

Between the early 1960s and the early 1980s, you held your breath, kept a sharp watch, and walked gingerly when pushing into the waist-high tangle of vines that draped across the low, secondary-succession vegetation repopulating what were old fields, pastures, and roadsides well before the ACG came into existence. Covered with concealing leaves in the rainy season, these tangles held an occasional and very unpleasant surprise, a nest of "lengua de vaca" (cow's tongue) wasps, *Polistes instabilis* (Vespidae). The wasps reacted aggressively to big mammals approaching their nests. These rust-red and yellow wasps had a characteristic manner: They came forth in plentiful numbers, buzzing out from between the leaves somewhat slowly and horizontally, aimed directly at whatever intruding body part was close by, typically a face or an arm. They made you pay a price for your trespass.

Upon disturbance, *Cosmosoma cingulatum* launches into instant flight just as you expect of a diurnally active insect. In flight, the moth has nearly the same wingspan as *P. instabilis.* In the dappled shade, it is a very similar color, and it flies with the same horizontal, buzzing, flat trajectory. It does not hesitate to fly straight at you, but *C. cingulatum* does not sting. In those early formative days of the ACG (Janzen 2000), this moth and an encounter with its wasp-like behavior was extremely common. In its abundance, for reasons quite unexplorable today, the adults aggregated and perched in the daytime within the same vine tangles used by the nesting *P. instabilis.* When that vine tangle was bumped, the moths exploded into the air exactly as would the wasps from a large nest. Coincidence? Not likely. The wasp mimicry is so evident at the level of the individual moth, with its clear wings, wasp-shaped body, wasp coloration, and wasp behavior. Furthermore, the ruse may be considered in a social context: The moths in aggregation create the illusion of a swarm of angry wasps.

Currently, *C. cingulatum* is rare in the ACG dry forest—one here and one there—and even more rare to appear at a light trap (they flee readily in the daylight, but also fly at night to visit flowers, mate, and oviposit). Where have they gone? The super-abundance may have been simply one of the inexplicable population explosions followed by a severe decline, a characteristic of many species of ACG dry-forest Lepidoptera (Janzen 1981). The decline may also be a direct outcome of the changing plant communities in response to the abatement of human disturbance and fire, perhaps exaggerated by the warming and drying that is happening to ACG dry forest as global climate change comes upon us. A fascinating question remains: Did *C. cingulatum* numbers decline as a response to a decline in the wasp model that it mimicked? The density of *P. instabilis* nests today in ACG dry forest is not even 10 percent of what it was two decades ago. No one today pushes into a vine tangle with mental trembling, hoping that it does not erupt with violent warriors. Part of the answer to our seemingly rhetorical question may be revealed by *Cosmosoma teuthras* (#24), an extraordinarily similar arctiid.

ADULT VOUCHER: 03-SRNP-15467; JCM
CATERPILLAR VOUCHER: 89-SRNP-386; DHJ

24. *COSMOSOMA TEUTHRAS* – ARCTIIDAE

Look from right to left and back again in the gallery. Did we make a mistake and put up two plates of the same species, one a worn *Cosmosoma teuthras* and the other a less rubbed *C. teuthras?* For two decades this pair fooled the inventory project. Although *C. cingulatum* has a black, scale-free stripe down the middle of the thorax, and *C. teuthras* has the same area covered with deep red scales, it is common for the dorsal center thorax of moths to have the scales worn off, partly obliterating its central color patch. Even the drawers of specimens in the National Museum of Natural History at the Smithsonian Institution contain mixtures of these two species that are so easily viewed as one. In 2004 Bernardo Espinoza noticed that two names had been applied to what seemed to be one animal. He examined the male genitalia and realized that the taxonomists of previous centuries had been quite correct to describe two different species. Now it is clear that the dark area in the middle of the thorax of adult *C. cingulatum* is not simply rubbed thoracic cuticle showing through, but a species-diagnostic trait.

Once we realized that there were two species, our interest turned to mapping their distribution. All the specimens in the INBio collection (see Gámez 1991, 1999) were examined that had been accumulated by parataxonomists "mothing" all over the country. It was instantly evident that *C. teuthras* is a species of the rain forest, especially the lowlands, whereas *C. cingulatum* is a denizen of dry forest and intermediate-elevation rain forest. Today, as the ACG dry forest dries and heats, it is *C. teuthras* that has disappeared almost entirely while *C. cingulatum* hangs on in low numbers. But what of the large swarms described in the account of *C. cingulatum?* They were a mix of the two species (both have caterpillars feeding on the same sapindaceous vines), and the ACG dry forest was a lot wetter then, both in the rainy season and in the dry season, which was shorter and less intense. The two species still co-occur in ACG rain forest and the intergrade between dry forest and rain forest, but the days are probably gone when swarms of newly eclosed adults of both species roosted together in a vine tangle in the second month of the rainy season.

What of the *P. instabilis* wasps they mimic? The annual migrations of huge numbers of these wasps out of ACG dry forest into the cold moist clouds of the adjacent volcano tops (Hunt et al. 1999) is now becoming a thing of the past. The heat from the lowland is rising up the volcanoes (e.g., Pounds et al. 2006) and displacing the colder conditions that prevailed upslope. The wasps' mountain refuge has allowed them to escape the six-month dry season in the dry-forest lowland, a time and place their prey caterpillars were almost nonexistent. Present unsuitable conditions for passing the bad time of year (dry season) will result in fewer individuals during the good time of year (rainy season). In turn, fewer wasps will exist to repopulate the next season's generations when caterpillar prey are plentiful. Hence the population spirals downward so that today a *P. instabilis* nest in ACG dry forest is a rare and fascinating sight, rather than a well-remembered horror.

ADULT VOUCHER: 83-SRNP-1221; JCM
CATERPILLAR VOUCHER: 83-SRNP-1221; DHJ

25. *HISTORIS ODIUS* – NYMPHALIDAE

25

Ranging from south Texas (albeit doubtful sightings) to well into South America, *Historis odius* is one of the large, strong, fruit-feeding nymphalids that seem to be about everywhere. Put out a serving of fermenting mangos or bananas and these powerful butterflies will appear, whether in the driest or the wettest ACG habitats. This butterfly, along with its close relative *Historis acheronta*, which feeds on rotting fruits still hanging in tree crowns as well as fallen fruits and damaged, sap-oozing tree trunks, are often accompanied by other powerful, large nymphalids such as *Prepona* and *Archaeoprepona.* All of these butterflies aggressively bash each other with the strong anterior margin of the front wings as they push and shove to get to the best feeding point. *Historis odius* never seems to visit flowers, though it is somewhat of a puzzle as to what they have against flower nectar. Maybe it is a lack of alcohol moieties on certain molecules?

The large, spiny, but harmless *H. odius* caterpillars are excellent Batesian mimics of fiercely urticating, hemileucine saturniid caterpillars and feed on the large leaves of various species of *Cecropia* (Cecropiaceae), but seem to ignore the other ACG Cecropiaceae. In ACG dry forest they pass through one to two generations during the first half of the six-month rainy season. Then most of the adults leave, probably flying to the rain forest in the east. In the rain forest, *H. odius* caterpillars can be found year-round, but never at the high density that can nearly defoliate isolated *Cecropia* trees in the first month of the rainy season in ACG dry forest. It is probable that the lack of *Historis* caterpillars on the few leaves sustained by *Cecropia* during the dry-forest dry season is a result of the adults being reproductively "turned off," though it is not clear what selected for avoidance of a food plant during the dry season.

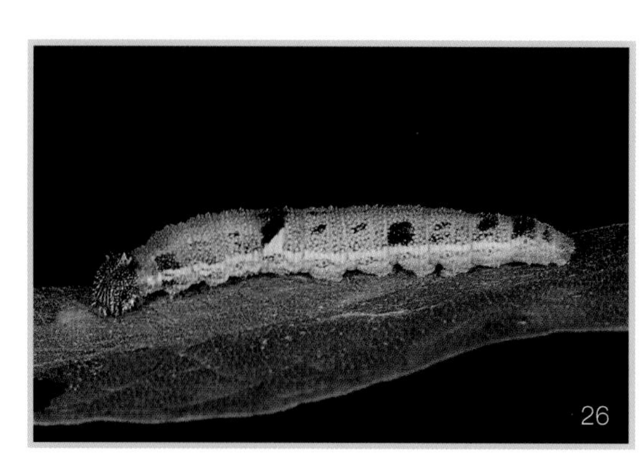
26

During the long dry season, adults of both species of *Historis* are present in ACG dry forest, though just about the only way to confirm their presence is by using binoculars to observe the woody, fermenting fruits of *Cassia grandis* hanging high in the canopy. A more convenient approach is to place fruit traps at ground level baited with fermenting fruit. The addition of a bit of spit and beer may improve the bait. Since *Cecropia* is primarily a human-associated invader of roadsides in ACG dry forest (much rarer or absent from natural disturbance sites), it is easy to wonder if the common presence of both species of *Historis* today is not an artifact that will disappear as the ACG transforms back to old-growth forest over the next decades and centuries.

ADULT VOUCHER: 05-SRNP-32305; JCM
CATERPILLAR VOUCHER: 00-SRNP-15841; DHJ

26. *ANAEA AIDEA* – NYMPHALIDAE

If you are from the south central United States, you may well know the goatweed butterfly, *Anaea andria,* the northernmost look-alike for *Anaea aidea,* which is yet another of the dead-leaf butterflies so frequently encountered at rotting fruit and tree wounds, never flowers, in the ACG. Breeding populations of *A. aidea* extend up into northern Mexico and even into Arizona and Texas on occasion. It is not difficult to imagine the origin of the North American extratropical *A. andria* as a northern evolutionary segregate of the tropical *A. aidea.* It could easily have been partitioned off by one of the glaciation events, evolved in isolation, and is now a distinct species. In this context, it is not surprising to find that *A. aidea* is the environmentally toughest of the many ACG members of this group of *Memphis-Anaea-Fountainea.* Caterpillars and adults of *A. aidea* are steadfastly present in ACG dry forest wherever their food plants, *Croton,* occur. This is true even for the Santa Elena Peninsula, a desolate place in the western region of the ACG that is swept by the wind and blasted by the sun. As is common for insects that do very well in the harsh environment of the dry forest, *A. aidea* does not invade the adjacent rain forest or cloud forest, even though two or three species of *Croton* are heavily used by many species of *Memphis* and *Fountainea* caterpillars. Why *A. aidea* does not colonize habitat beyond the dry forest is truly a puzzle. This butterfly has not even followed the invasion of the rain forest by dry-forest pastures and fields, as has been the case with many other species of ACG dry-forest Lepidoptera.

The first instar of *A. aidea* feeds and perches on the leaf tip, as do other charaxine nymphalids, appearing to be a bit of trash, an appearance that is sometimes augmented by bits of leaf silked in place to enhance the effect. As it reaches the fourth instar, the caterpillar constructs a tube out of the leaf blade in which it lives, venturing out at night to feed on other leaf blades. In contrast to many leaf-rolling moth larvae, it neither eats the leaves that make up its living quarters nor allows its fecal pellets to accumulate inside the tube. Such an abode may well protect against birds, caterpillar-killing wasps such as *Polistes instabilis* (Hunt et al. 1999), bats (Kalko and Kalko 2006), spiders, and other such carnivores, but it is certainly no absolute protection against at least one species of braconid wasp (*Cotesia*) and five species of tachinid flies. These parasitoids kill over 10 percent of *A. aidea* caterpillars in the ACG.

ADULT VOUCHER: 91-SRNP-2151; JCM
CATERPILLAR VOUCHER: 05-SRNP-59158; JCM

27. *DYSSCHEMA VIUDA* – ARCTIIDAE

It is impossible to escape the supposition that the describer of *Dysschema viuda* knew this moth's caterpillar, though there is no mention of it in the original description (Schaus 1910). "Viuda" means "widow" in Spanish. The caterpillar of *D. viuda*, one of the largest arctiid caterpillars in the ACG, ostentatiously and diurnally clambers over many plant species, its black and gray-white, densely hairy body, perhaps conveying to William Schaus a veiled and somber existence. A different thought on Schaus's intent has been offered by Vitor Becker, who suspects that Schaus called this moth a widow because at the time of description he had only a female specimen; no males had been caught. We may never know what Schaus was thinking.

Caterpillars of *D. viuda* are quite the generalists for food plants (at least five plant families to date, but probably many more to come) and for where it rests. This restless caterpillar seems to be constantly on the move. Capturing the caterpillar photo we show here was nothing less than an ordeal. Not only is it restless, but if touched or jostled when it is calmly feeding, it readily drops off the plant to race off in another direction. A long list of caterpillar food plants is common among the Arctiidae, but *D. viuda* carries to an extreme the restlessness that often goes with it.

27

Dysschema viuda is a nocturnal rain-forest moth, with both sexes arriving occasionally at lights. The females, with much more bright yellow than the male we show, are also active in the daytime, looking like some kind of huge and fast-flying heliconiine butterfly. It is not clear why *D. viuda* does not venture into ACG dry forest—some of its caterpillar food plants or close relatives in the Solanaceae do occur there. When the brightly colored moth is grabbed by a collector, it emits a strong odor that is presumably part of its defense, though it is likely that its chemical defense consists of much more than a bad smell. All members of the genus *Dysschema* are brightly colored and generally assumed to be models in Mullerian mimicry complexes. This is probably an accurate assumption based on the nasty chemicals sequestered by the caterpillar from the larval food plants (Asteraceae, Araliaceae, Piperaceae, Solanaceae, etc.). However, there also remains the possibility that all the caterpillars of this arctiid subfamily, Pericopinae, manufacture their own nasty chemical defenses.

ADULT VOUCHER: 04-SRNP-34244; JCM
CATERPILLAR VOUCHER: 04-SRNP-33231; JCM

28

28. *DANAUS PLEXIPPUS* – NYMPHALIDAE

This could hardly be a photographic gallery of butterfly portraits representing the Neotropics without including the monarch, *Danaus plexippus,* the only tropical butterfly that the vast majority of North Americans will ever see on the wing. Tropical, you say? Yes, *D. plexippus* is unambiguously a tropical butterfly. Furthermore, all of its danaid relatives are tropical butterflies. ACG monarchs are smaller, with slightly shorter forewings, and they are not part of the population that migrates annually between Mexico and the northern reaches of North America. Yet their biology in the ACG gives some clues as to how this migrational behavior evolved.

A breeding population of *D. plexippus* with overlapping generations is present throughout the year in ACG rain-forest pastures, fields, and occasionally naturally disturbed sites. Human disturbance has been so pervasive it is truly impossible to know how this species lived in pre-European and even pre-human ACG rain forest. The vividly aposematic caterpillars, rich in toxic cardiac glycosides, are easy to find on herbaceous Asclepiadaceae. Its breeding and feeding activities extend even up to the lower cloud forest in pastures, though this may be entirely the result of humans warming and drying the habitat with their forest clearing. When the six-month dry season breaks about 15 May, the greening and wetting of the dry forest brings with it an invasion of *D. plexippus* from the rain-forest side of the ACG. Some adults also move out of the densely shady, moist habitats they have been occupying along watercourses, linear oases through the leafless dry forest in the dry season. They then pass through two to four generations on the herbaceous Asclepiadaceae of ACG dry forest. As the next dry season arrives, the strong, flying adults gradually retreat upwind, back into the deeper rain-forest parts of the ACG and down into wet river courses. Although this pattern does not contain the behavior of mass aggregations and dormancy in the cold mists of upper elevations, as is the case with the migrating North American population of *D. plexippus,* it is easy to imagine how this behavior could be added into the ACG monarch traits. Whatever the case, *D. plexippus* is unambiguously a nomadic to migrant species within the ACG. Its behavior is clearly congruent with the evolution of behavior in more northern populations in response to glacial retreat. The monarch followed the retreating ice and spread across North America.

ADULT VOUCHER: 92-SRNP-3514; JCM
CATERPILLAR VOUCHER: 98-SRNP-7073; DHJ

29. *XYLOPHANES PORCUS* – SPHINGIDAE

Xylophanes porcus is a species everyone knows. This is the medium-sized sphingid with wings and a body that are dull, drab olive, and nearly patternless. It comes to light traps just about everywhere except at the highest elevations. Hundreds of them are pinned and spread in private collections, but usually not more than a few specimens per collection because they are neither prized for their beauty nor rare in nature. They look superficially like the adults of *Xylophanes juanita* (like *X. porcus* but with stubby, shortened wings), *Xylophanes germen* (crenulated wing margins and a slight pattern on the front wings), and *Xylophanes hannemanni* (much more falcate forewings, with a blurry, shiny area in the front wings).

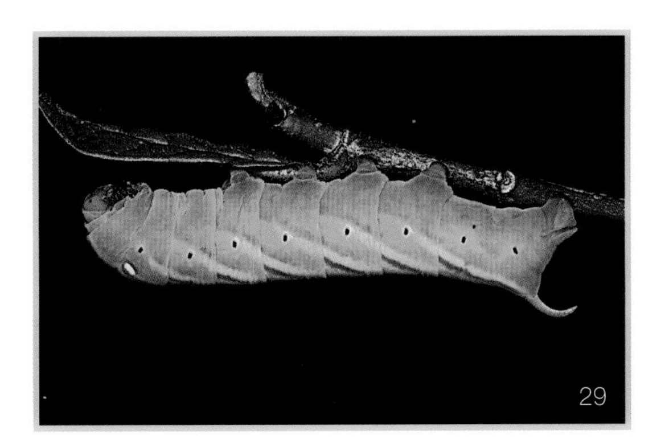
29

Once again, DNA barcoding provides new information. As a common moth, *X. porcus* is mostly ignored by ecologists and taxonomists because it is well known. It turns out that the DNA barcode sequences group this "species" into three distinct clusters, one from specimens reared largely in ACG dry forest and two others from specimens reared largely in ACG rain forest. To date we can find no morphological difference between what are undoubtedly three species living side by side in the ACG.

So how will the real *X. porcus* reveal itself? Unfortunately, there is no quick way to know which of the DNA barcode sequences actually matches the holotype of *X. porcus,* a specimen long ago deposited in some extratropical museum. The obvious quick answer is to compare a barcode DNA sequence from the holotype with that of the three biological entities in the ACG. But that will require finding the holotype and getting permission to remove a leg for the DNA analysis. A positive match would suggest that we could then identify specimens that are the true *X. porcus.* However, a negative finding (poor DNA match) does not necessarily confer "different species" status, because one of the ACG entities might represent a population that is part of a cline extending from wherever the type was collected to that population in the ACG. In other words, to really understand the limits of the species, it is important to study various populations across a representative geographic range, something we did for *Xylophanes libya.*

When we mapped *X. libya,* which showed a pattern of many entities based on barcode data, we found that there were also three *X. libya* groupings in Costa Rica: one in the dry forest, another at intermediate-elevation rain forest on both the Atlantic and Pacific side, and a third in the deep lowland rain forest of the Caribbean. But all three overlap in some places, so it will be necessary to barcode the type specimen. So it is with *X. porcus*—geographical location is not a diagnostic trait. For the moment, there is no choice but to use the name *X. porcus* and anticipate that a new name may occur as this taxonomic tangle is unraveled. On the ACG inventory Web site, the ACG species identified by the barcoding are known by the interim names *Xylophanes* JANZEN03 and *Xylophanes* JANZEN04.

ADULT VOUCHER: 03-SRNP-16081; JCM
CATERPILLAR VOUCHER: 04-SRNP-12011; JCM

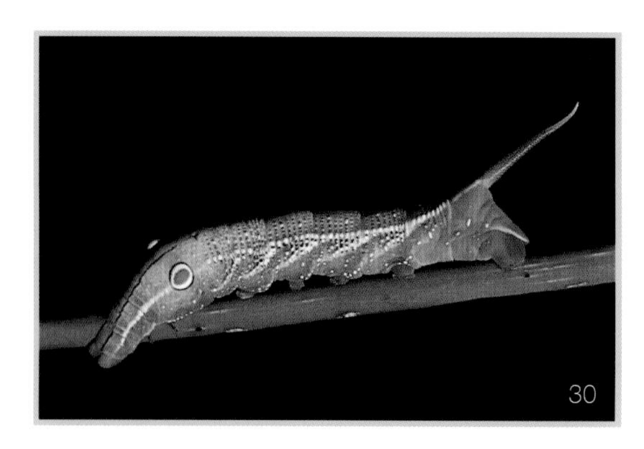

30. *XYLOPHANES CROTONIS* – SPHINGIDAE

Xylophanes crotonis is the well-known look-alike to the most recently described new species of sphingid in Costa Rica, *Xylophanes letiranti* (Vaglia and Haxaire 2003). These species are essentially identical, but *X. letiranti* has a sprinkling of black scales dotting the yellow patches in the hindwings. Or perhaps we should say it has substantially more black scales dotting the yellow patches of the hindwings, since *X. crotonis* has a few of these scales itself. There are also very slight differences in the genitalia of the two species. To date, only *X. crotonis* is known from the ACG, but since *X. letiranti* occurs at Monteverde, only 150 kilometers to the south, and at about the same elevation as the upper parts of the Cordillera Guanacaste, it won't be surprising to find it someday in the ACG. It seems very likely that *X. letiranti* is no new arrival on the scene, but rather that it has simply been hiding within the mental image we have formed for the long-known *X. crotonis* adults.

Adult *X. crotonis,* based on its responsiveness to lights and on the abundance of caterpillars on food plants, is the upper-elevation, common *Xylophanes,* along with the extremely rare *X. godmani.* High on Volcán Cacao or high in the central Talamanacas of Costa Rica, *X. crotonis* is the one resident *Xylophanes* that can be relied on to arrive at a light trap. Others, such as *X. chiron, X. ceratomioides,* and *X. porcus,* occasionally appear at high-elevation lights, but these (and many others that appear at high mountain passes) are low- to middle-elevation species migrating from one side of Costa Rica to the other, passing over high mountain passes in the process. Moving down to middle or low elevations, the number of species of *Xylophanes* increases to ten to twenty-three species, depending on the site. All of the Costa Rican *Xylophanes,* except *X. letiranti,* have been found in the ACG, and probably all breed there.

Caterpillar hunters have dubbed *X. crotonis* with a colloquial name, "blue bulgy eyes," and a glance at the somewhat comical caterpillar makes it obvious why. Two species, *X. crotonis* and *X. germen,* whose bulging false eyes are bright red, are the only two Costa Rican sphingids with truly bulging false eyespots. There appears to be no similarity of the adults of these two species, which leads to the inference that they have independently evolved. This exceptional exaggeration of the false eyes may be a sort of super startle stimulus. The mid-elevation to lowland rain-forest *X. titana,* the adult of which looks like a smaller and paler version of *X. crotonis,* is easy to recognize as a close relative to *X. crotonis.* The caterpillar of *X. titana* not only has a general body pattern very similar to that of *X. crotonis,* it also has blue false eyespots, though they only bulge mildly instead of prominently.

ADULT VOUCHER: 99-SRNP-426; JCM
CATERPILLAR VOUCHER: 98-SRNP-3157; DHJ

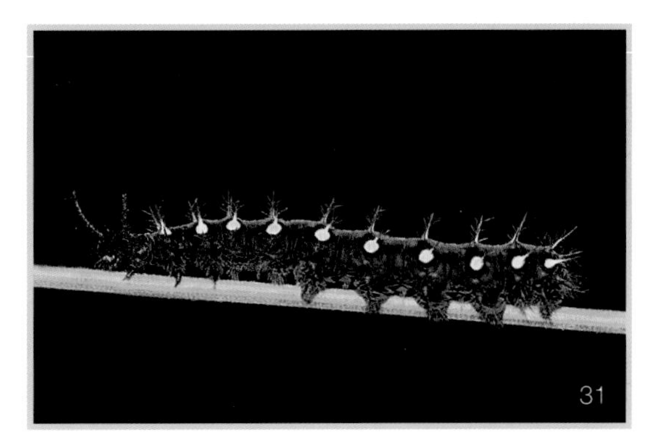
31

31. *SIPROETA EPAPHUS* – NYMPHALIDAE

The jet-black caterpillar of *Siproeta epaphus,* with its bright yellow dots highlighting what appear to be fiercely urticating spines, is probably a combination of mimic and model. The spines do not urticate, but look like they should. One wonders how much evolutionary change in physiology and morphology it would take to modify those spines into urticating structures, an evolutionary trick that other spiny mimetic nymphalids have not achieved either. The black body, however, may well signal a body filled with nasty chemicals sequestered from the acanthaceous food plants of *Blechnum, Justicia,* and others, as does the black tubercule-laden body of *Battus* (Papilionidae) caterpillars feeding on Aristolochiaceae.

Viewed in the collector's drawer of butterflies, the large orange, black, and white *S. epaphus* seems to be a form and color all to itself, with no mimics and no look-alikes. But once again, back off and blur your eyes a bit, and now you believe you are looking at yet another *Adelpha* color pattern: dark background, a large amount of orange in the forewings, and white markings down the wing. Visually, though not taxonomically, *S. epaphus* is a member of the many tens of *Adelpha*-like species in ACG rain forest. Whether it is a Mullerian or Batesian mimic remains to be seen, but it is certainly no isolated accident of color.

At the edges of the ACG rain forest, *S. epaphus* is accompanied by the dry-forest butterfly *Siproeta stelenes,* which on first glance appears to be wildly different, with large, jade-green patches on dark brown wings. But the caterpillar of *S. stelenes* is also black and spiny (advertising an urtication that it does not have), and also feeds ostentatiously on acanthaceous herbs. Along with *S. stelenes* flies *Philaethria dido,* an unambiguous heliconiine mimic of *S. stelenes,* or vice versa. In mid–rainy season on the interface between ACG dry forest and rain forest, both *S. epaphus* and *S. stelenes* adults can be so common that they seem to be everywhere. The small, insectivorous wrens and flycatchers in their vicinity pay absolutely no attention to either species. All these signals point to both species being aposematic (and perhaps Mullerian models) rather than Batesian mimics. However, the most important function of such conjecture is to emphasize the need for yet more feeding experiments.

ADULT VOUCHER: 02-SRNP-27752; JCM
CATERPILLAR VOUCHER: 04-SRNP-46880; JCM

32. *ADELPHA BASILOIDES* – NYMPHALIDAE

If there was ever a mimetic mess in the ACG it is among the adults of the species of *Adelpha.* Some fourteen of the seventeen ACG *Adelpha* species reared to date look like one another, especially when flying, with the exception of the two cloud-forest species, *A. tracta* and *A. demialba.* These two deviate from the lowland *Adelpha* by displaying a common shift in appearance. The basic *Adelpha* adult color pattern, here portrayed by *Adelpha basiloides,* plays out in seemingly infinite minor variation in more than 100 other Neotropical species and even with a few species in North America (Willmott 2003). Among all of the species of *Adelpha* in the ACG, we know of only one that is common, *A. basiloides,* which happens to be distinctive for a single trait in the forewing. No doubts in this identification. Our marker separating it from all the others was the split white cell at the top of the column of multiple white cells lined up on the dorsum of the forewing. We were happy with this diagnostic trait until we assessed its DNA barcodes. What we and everyone else knew as *A. basiloides* is clearly two entities in the ACG. One of these entities is a dry-forest species we have temporarily labeled *Adelpha* JANZEN01. The other entity, temporarily known as *Adelpha* JANZEN02, appears to be primarily occupying the interface between rain forest and dry forest. The two entities can be found together in the zone of overlap between the two major ACG ecosystems where their caterpillars feed on the same species of plants.

32

What does the orange and white pattern on a dark brown background mean in the life of an *Adelpha?* It is easy to suspect that it is an aposematic coloration because these butterflies fly, float, flap, and perch in the vicinity of fly-catching birds that certainly should have the ability to catch *Adelpha,* but show no interest. Also, at least for many species, the cryptic larvae feed on rubiaceous plants that contain enough potentially toxic alkaloids that they could be sequestered as defensive chemicals. This is yet one more case where only detailed, and perhaps somewhat unpleasant (for the test animals) experiments could determine if they are models, mimics, or both. Interestingly, there are numerous other nymphalids and riodinids that bear the same color pattern (see images in DeVries 1987, 1997), apparently mimicking *Adelpha* and forming a yet larger mimetic complex. Among the more dramatic are the several species of *Doxocopa*, such as the female of *D. pavon* and both sexes of *D. laure*, which belong to a different nymphalid subfamily from *Adelpha*, and yet contain species that are extremely similar to *Adelpha*.

ADULT VOUCHER: 81-SRNP-394; JCM
CATERPILLAR VOUCHER: 03-SRNP-21218; DHJ

33. *MORPHO THESEUS* – NYMPHALIDAE

When we say "morpho," everybody thinks, "big, blue, beautiful butterfly flopping along the roadside anywhere and everywhere." In fact, the big, blue, beautiful *Morpho peleides* is in ACG cloud forest, rain forest, and dry forest. Two other species of blue *Morpho* inhabit only the rain forest, *M. amathonte* and *M. granadensis*. But there are other species of *Morpho*, species that are not blue. The pearly white-winged *M. polyphemus* (Miller et al. 2006) flies in the cloud-forest canopy in August, a short flight season for this univoltine species. Then there are the brown and beige *Morpho* with pearly wings. In the ACG, this would be *M. theseus*, whose underside we portray in the gallery.

The adult *M. theseus* is enigmatic, at least in the ACG. Whereas the white morpho can be found reliably on the tops of volcanoes in August, and the blue morphos are sort of everywhere all the time, we did not meet our first brown morphos until 1991. Suddenly in March they were abundant, flying up and down all the trails on Volcán Cacao at an elevation of 1,000–1,400 meters, and suddenly they were gone, not to be seen again for a decade. Then we met them again, this time as caterpillars on what we now believe to be their sole ACG food plant, the large perennial vine *Abuta panamensis* (Menispermaceae), at an intermediate elevation in the rain forest. Where were they in between our sightings, and where are they now? Recently their food plants are largely or entirely free of the very distinctive caterpillars. Perhaps the population moves from place to place. If this is the case, it represents a different lifestyle from the blue morphos, whose caterpillars feed on a variety of species of Fabaceae and even some other distant plant families. Now is the time for tiny transmitters to be glued to the back of a brown *M. theseus* so it can be followed for many kilometers to who knows where.

Eggs of *M. theseus* are laid ten to thirty per batch. The early instars feed side by side on the large leaves of *Abuta*. When the caterpillars reach the last instar, they are solitary, scattered over the food plant, but still ostentatious. Not only does *M. theseus* differ from other morphos by being gregarious and very host specific, they are parasitized by only a single, undescribed species of *Hyphantrophaga*. This tachinid fly attacks only *M. theseus* caterpillars. Where is this fly when the caterpillars are not present? Maybe it, too, moves long distances? The caterpillars of the other four species of ACG *Morpho* do not provide the answer. They are attacked by their own distinctive parasitoid, a different and yet undescribed species of *Hyphantrophaga.*

ADULT VOUCHER: 02-SRNP-23688; JCM
CATERPILLAR VOUCHER: 02-SRNP-29527; DHJ

34. *MORPHO AMATHONTE* – NYMPHALIDAE

We cannot compose a book about Costa Rican Lepidoptera without at least one image of a bright, iridescent blue *Morpho* butterfly. The usual species illustrated in tourist brochures is *Morpho peleides,* easily reared by butterfly farms, having a strong, wide, black band around the margins of the wings (Miller et al. 2006). Also, *M. peleides* is the species so commonly encountered flying at eye level down forest trails and along roadsides in the rain forest, dry forest, and cloud forest. The species we use to represent the iridescent blue-winged group in the genus is *Morpho amathonte.* This species is exclusively an occupant of the ACG rain forest, not venturing up the slopes of the volcanoes and into cloud forest, the home for the white-winged *M. polyphemus* (Miller et al. 2006). Neither does it venture into dry forest, even though some of its caterpillar food plants, Fabaceae, Dichapetalaceae, and even the occasional palm, are found in these other two ecosystems. Males (our portrait) of *M. amathonte* are distinctive, having a black border only at the wing tip and along the forewing costa. Females do have a wide, brown-black, and spotted border to the blue field (see http://janzen.sas.upenn.edu). Both sexes, with wings larger than those of *M. peleides,* have a swooping, flopping flight and often go as high as ten to thirty meters above the ground as they course up and down road cuts through the forest.

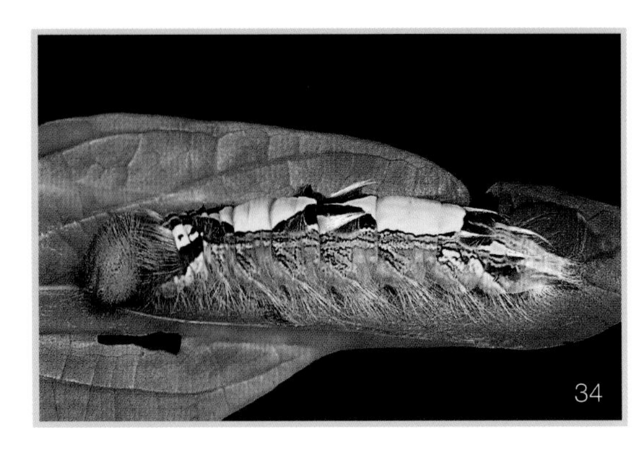

All of the ACG *Morpho—M. amathonte, M. peleides, M. granatensis, M. polyphemus,* and *M. theseus*—show a constellation of false eyespots on their underside (see *M. theseus,* #33). The protective function of these eyespots is clearly not engaged when the butterfly is in flight, but most likely when roosting on foliage, or when perched on rotting fallen fruit on the forest floor. They, like many saturniines and brassolines with similar false eyespots (see *Opsiphanes cassina,* #35 and *Caligo atreus,* #36), offer a super stimulus on close view that says "you are being looked at"—a stimulus that if not immediately responded to by instant flight could make a meal of the observer were the eyes indeed real.

Caterpillars of *M. amathonte* are morphologically similar to those of *M. peleides* and share a number of food-plant species in the ACG rain forest. The dorsal view of a last-instar yellow morph of *M. peleides* will reveal a number of small details that allows it to be distinguished. The easiest key character is the presence of two central dots, small and yellow against a dark background on *M. peleides,* absent in *M. amathonte.* The apple-green pupae are also very similar in form, but the pupa of

M. amathonte sports a brilliant white, horizontal belt across its middle section (see page 11), whereas that of *M. peleides* is immaculate and a plain, uniform green.

ADULT VOUCHER: 99-SRNP-5893; JCM
CATERPILLAR VOUCHER: 03-SRNP-6650; JCM

35. *OPSIPHANES CASSINA* – NYMPHALIDAE

It has been a long day, the sun has just dropped below the horizon, and you settle back into your chair in front of the ACG dry-forest house to enjoy the famously short tropical twilight. There is a blurry, dark, and lightning-fast butterfly sallying out from its perch, a branch tip, flying off frantically a few tens of meters, and then returning. A minute later, it repeats the behavior. The last of the typically diurnal butterflies quit flying for the day about an hour ago. You have now met *Opsiphanes cassina,* one of the handful of ACG butterflies that fly only at dusk. It appears to have an amazingly short flight period, perched motionless all day under a large leaf and apparently not active in the deep of night. It does not come to lights or fruit baits at night, and it is never seen in a flashlight beam. During a very brief period at dusk, however, it does come to fermenting fruit, probing with a powerful tongue equal to that of any fruit-feeding, large, charaxine nymphalid such as *Archaeoprepona, Prepona,* and *Agrias.*

35

Individuals of *O. cassina* are everywhere in ACG dry forest that sports its caterpillars' food-plant palms, *Bactris major, Bactris guineensis,* and *Acrocomia aculeata.* The former two are native and probably are the "real" food plants for this species. *Acrocomia aculeata* may be considered an anthropogenic species due to its megafauna fruit (Janzen and Martin 1982). Before the Spaniards arrived with their livestock, the plant was probably either absent from Costa Rica or primarily associated with human activities, owing to its highly edible fruit. Cattle found the highly nutritious fruit pulp much to their liking, but they could not break the nut-like seed and either spit them out when chewing their cud or passed them through in their dung. They dispersed the seeds into the full sun of pastures, the ideal habitat for *A. aculeata* (or *coyol,* as it is called). The seedlings must have full sun to become a mature plant. With the advent of cattle grazing and the expansion of pastures came the demise of the forest and forest edge–loving *Bactris.* A scarcity of resources, namely the favored food plants, resulted in *O. cassina* caterpillars moving from *Bactris* to *Acrocomia* as their primary food plant. Today, as the cattle industry shrinks, taking *Acrocomia* with it, and with *Bactris* slow to return, the *O. cassina* population is headed for a bottleneck in Guanacaste Province.

There are several possible survival scenarios for *O. cassina.* First, the butterfly has been quick to adopt introduced ornamental palms, such as *Cocos nucifera* and *Chrysalidocarpus lutescens*, as quality food for its caterpillars. Second, as the cattle industry spread into Costa Rican rain forest, it took the *Acrocomia* nuts with it, and dotted the plants across that insolated wet landscape. *Opsiphanes cassina* moved right along with it. Such refuges might serve as islands from which it ventures to use the foliage of at least five of the rain-forest palms (and perhaps to use all of them, given the right circumstances), though presently it conspicuously avoids relatively intact rain forest. After over 500 *Opsiphanes* caterpillar field collections, all from a total of a dozen species of native ACG rain-forest palms, only five (1 percent) were cases where the caterpillar was developing on palm foliage in true rain forest.

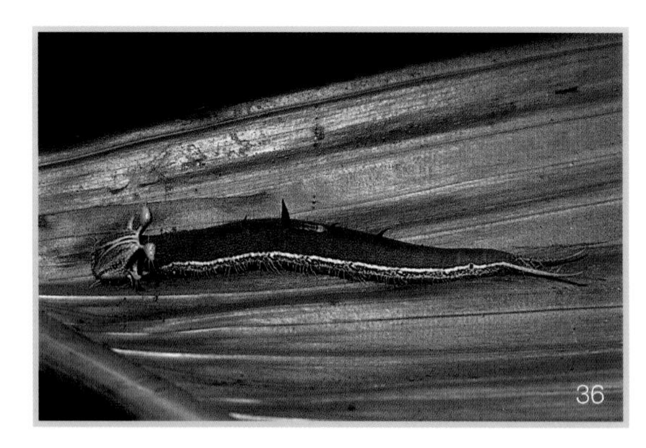

ADULT VOUCHER: 03-SRNP-14224; JCM
CATERPILLAR VOUCHER: 03-SRNP-1951; DHJ

36. *CALIGO ATREUS* – NYMPHALIDAE

When Janzen was fourteen years old, in 1953, he took his first trip to Mexico. One day while he was standing at what was then the forest edge just outside of Tamazunchale (today a degraded pasture), a monster butterfly emerged from the forest, flapped frantically and erratically over his head, and then started to fly away. That was taken as a challenge and it was a definite mistake by the butterfly, given a teenager's quick reflexes. Dan's father took a picture of his proud son holding a *Caligo atreus*. Dan's attention was first drawn to the garishly yellow and blue upper side of the wings, a color pattern that would draw the eyes of any pursuit predator. Then he recalls wondering about the gorgeous "owl eyes" on the underside of the wing. These eyes give all species in the genus the well-known common name "owl butterflies."

The large, false eyes of the over twenty species of *Caligo* distract from the detail of the fine ripple reticulation and smaller eyespots that also adorn the underside. Yet those finely detailed markings contain valuable information that can be used to distinguish among the species as reliably, if not more so, than any other easily visible trait. We use those traits to identify the adults of the five species that have been reared in the ACG to date: *C. telamonius* (formerly called *C. memnon*), *C. eurilochus, C. oedipus, C. illionius,* and *C. atreus.*

Caterpillars of *C. atreus* are the champion generalists among the ACG species of *Caligo.* The solitary (from the time of oviposition) caterpillars have been found feeding on five species of Zingiberaceae, one species of Musaceae (introduced), seven species of Marantaceae, eight species of Heliconiaceae, one species of Cyperaceae, seven species of Costaceae, one Cyclanthaceae, one Bromeliaceae, and nine species of Araceae (palms). Nothing in the DNA barcode suggests that *C. atreus* is a complex of cryptic species making use of food plants in so many higher plant families. However, it is notable that all of these plants are monocots—no caterpillar of a species in the Brassolinae has ever been found feeding on a broad-leafed (dicot) plant. If there ever was a candidate species among the Brassolinae that might be able to switch food plants from monocots to dicots, *C. atreus* is it. It is as if the female operates on the dictum, "If the leaf is broad and long, lay an egg on it." The diversity in food plants may also be part of the secret of how it can range from near to the northern end of the tropics in western Mexico all the way to southern tropical South America. Yet, even with this obvious great breadth of food plants, it avoids the dry forest and occurs only in the rain-forest and lower cloud-forest portions of the ACG. Only *C. telamonius* maintains a breeding population in the ACG dry forest, a mere kilometer from the territory occupied by *C. atreus.*

ADULT VOUCHER: 03-SRNP-22394; JCM
CATERPILLAR VOUCHER: 01-SRNP-9052; DHJ

37

37. *PENTINA FLAMMANS* – THYRIDIDAE

This family of small moths, Thyrididae, is one of those Neotropical groups that are species-rich in the ACG and to the south, but have just a few species in North America. Most of the ACG species have been found by first encountering the caterpillars, but there are also many species that have come to light traps, and their caterpillars remain (as yet) undiscovered. What makes an ACG thyridid caterpillar immediately noticeable in the field, at least for those that have been discovered, is that instead of rolling up and constructing the leaf nest in which it hides in the manner that humans roll a cigarette or a map, the leaf is rolled as a sugar cone for ice cream or a dunce hat for the schoolroom corner. The bottom of the roll winds down to a point with no exit, whereas the top is wide open and somewhat blocked by the massive head and prothoracic shield of the chubby and sluggish caterpillar. The caterpillar is speckled with shield-like structures, shiny and hard, a sort of beginning medieval armor, presumably as part of its defense against parasitoids and predators.

The caterpillar of *Pentina flammans* conforms to this basic tropical thyridid pattern, and so rolls the margin of the very large leaves of an ACG rain-forest understory treelet, *Pentagonia donnell-smithii* (Rubiaceae). The bright golden-orange adults appear to be almost aposematic when spread and pinned, but when perched among the foliage in the forest understory, they look much like a chip of dry leaf stuck to another leaf or twig. When disturbed, they remain motionless, as you would expect of a dead plant part. This way of passing the day, avoiding death by predation, is general throughout ACG Thyrididae. It is striking to count the many ways a dead-leaf mimic has evolved.

Many such mimics have windows in the wings. In North America the thyridids have been baptized as the window-winged moths (Covell 1984). In the tropics, however, few species have windows in their wings.

ADULT VOUCHER: 99-SRNP-13754; JCM
CATERPILLAR VOUCHER: 99-SRNP-4402; DHJ

38. *OTHORENE VERANA* – SATURNIIDAE

In the early 1980s, when the ACG caterpillar inventory was a new project focusing on the dry forest of Sector Santa Rosa, the large caterpillars of Saturniidae fell quickly into the hands of field collectors and were successfully reared. Simultaneously, the species richness of adults collected at the light traps quickly tallied up in an asymptotic fashion. More nights and more years of sampling did not generate more species for the dry-forest saturniid list. We knew of twenty-nine species from the sampling at lights and twenty-eight species from the rearing of caterpillars. The single missing species on the caterpillar side of the ledger was *Othorene verana.* This species was common at the lights, though often confused by collectors with its look-alike *Othorene purpurascens* (Miller et al. 2006). So we knew *O. verana* had to be a common, large caterpillar, but it simply was not at hand, even under an assault of extensive, albeit haphazard, searching by an increasing number of people.

In 1984, in a state of frustration, we broke our own self-imposed cardinal rule, a rule that stated we would not explicitly search for any particular species. The reason for the rule is simple economics. Species-dedicated searching will seriously lower the inventory rate of discovery that may otherwise be expressed as caterpillar yield per U.S. tax dollar (grant funds spent). But even cardinal rules are written to be broken. Several plump, egg-filled, and fertilized females were caught at the lights. They deposited hundreds of eggs glued to the insides of their plastic bag enclosures. Six days later, the eggs hatched, and numerous bags, each containing five hungry first instars, were stuffed with several leaves of a given species of tree, three-hundred species of trees to be exact. It took only twenty-four hours for the caterpillars to reveal which of the three hundred test species was a suitable food plant. The caterpillars rejected all species of tree leaves except those of *Quercus oleoides* (Fagaceae), the one and only Central American oak commonly found in the lowland tropics. (Actually, *Q. oleoides* is really the same species as *Q. virginiana,* the Virginia live oak of the eastern and southern United States, which changes its name to *Q. oleoides* about where it crosses the border into Mexico). This evergreen oak once covered tens of thousands of hectares of ACG dry forest and still maintains what appears to be a healthy breeding population, which we now know supports a healthy breeding population of *O. verana.* Once we knew the food plant, a few days of intense searching led us into the upper canopy of the oak trees and the expected caterpillars of *O. verana*. It is likely that intense light-trapping where *Q. oleoides* occurs throughout lowland Pacific Mesoamerica and then across the Isthmus of Tehuantepec and north to Monterrey would find a continuous population of *O. verana*, a distribution already partly mapped by Lemaire (1988).

ADULT VOUCHER: 96-SRNP-5458.10; JCM
CATERPILLAR VOUCHER: 84-SRNP-140; DHJ

39. *MEMPHIS MORA* – NYMPHALIDAE

Those who study butterflies normally encounter *Memphis mora* feeding on fermenting bananas in a rain-forest butterfly trap, putting it with the gang of diurnal butterflies that include *Fountainea, Anaea, Prepona,* and *Archaeoprepona,* none of which visit flowers. But it is the caterpillar food-plant relationships of this species that hold a special place in the annals of ACG *Memphis* natural history. The ACG rain forest is densely populated with many tens of species of Lauraceae, *Ocotea, Nectandra, Licaria, Beilschmedia,* and *Persea.* The foliage of many of these plants is fed upon by caterpillars of the various species of *Memphis,* as well as by other charaxines. But *M. mora* feeds on the foliage of only one of them, *Ocotea cernea.* Out of 110 *Memphis* caterpillars found on *O. cernea,* every single one has been *M. mora,* and a *M. mora* caterpillar has never been found on any other species of food plant, despite having found more than 1,000 wild *Memphis* caterpillars. Obviously the *M. mora* caterpillar uses some type of phagostimulant that serves as the unique identifier for recognizing the proper food plant, but we have no idea as to which molecules those might be. The caterpillar is an excellent biochemist and a very good plant taxonomist.

There seem to be an overwhelming number of species of *Memphis* that are black with some blue iridescence on the dorsal surface of the wings (DeVries 1987), accompanied by a gray, brown and rust bark-colored pattern on the ventral surface of the wings (see our portrait of *Memphis proserpina,* #40). The males of *Memphis* tend to express the black and blue one way, and the females express the black and blue yet another way, generally with more blue. Also, it is common for the females to have larger and longer tails on the hindwing.

The male we portray has only a tiny bump on the margin of the hindwing. In ACG dry forest, there is an absolute look-alike, *Memphis moruus* (often incorrectly written *Memphis morvus*), diagnosed at a glance by its long tail where the bump is located on the *M. mora* wing. Interestingly, a single tail-less male was reared in Sector Santa Rosa, the ACG dry forest. Barcoding demonstrated fully that it was a specimen of *M. moruus,* as the dry-forest habitat indicated it should be, rather than *M. mora,* as the appearance indicated it should be. Not only are the adults extremely similar, but the caterpillars are almost identical as well. Both species sport the same red bar across their foreheads and the same yellow stripes on their faces. Just like its look-alike, *M. moruus* feeds on only one species of Lauraceae, but it is sort of cheating to make the comparison because the ACG dry forest has only one species of native Lauraceae, *Ocotea veraguensis.* These two butterflies are unambiguously extreme food-plant specialists.

ADULT VOUCHER: 04-SRNP-30117; JCM
CATERPILLAR VOUCHER: 03-SRNP-10173; DHJ

40. *MEMPHIS PROSERPINA* – NYMPHALIDAE

Memphis proserpina seems to be found in either one of two conditions; one, as a newly eclosed adult in a plastic bag hanging in the rearing barns; or two, as an intoxicated individual wobbling around the inside of a trap baited with fermenting bananas. The gorgeous adult reddish and fine, silver-etched underside we feature so conspicuously here does render the perched adults essentially invisible on tree bark, but that is not the reason we never see them in the ACG rain forest and cloud forest where they are common. Rather, they appear to pass almost their entire life high in the canopy. An exception is when they are attracted down to ground level by a fruit trap, whose odors, including that of alcohol and other microbial products, are rising up into the canopy.

Only three species of *Mollineda* (Monimiaceae) serve as caterpillar food plants: *M. costaricensis, M. pinchotiana,* and *M. viridiflora.* All are vine-like shrublets at ground level. As with other species of canopy-living, strong, flying nymphalids, the female plummets down from high above, bounces from shrub to shrub "tasting" with her feet, finds a future caterpillar food plant, lays an egg, and a few seconds after her descent she is again high up in the canopy. Were it not for the tracks left by caterpillars feeding for three weeks or more, her activities would be invisible to the observer.

The early instar resembles a fragment of damaged leaf and feeds on the leaf tip, as do other young charaxine caterpillars (see *Anaea aidea,* #26; and *Fountainea eurypyle,* #7). The later instar lives inside a silk-lined rolled leaf as a chubby black-bodied and white-dotted caterpillar that ventures out at night to feed before returning to its diurnal retreat. This lifestyle is common among the many species of ACG *Memphis,* as is the behavior of a rapid descent from high above to lay an egg. Because they are in the leaf roll in the daytime, it is difficult to grasp what selects for the distinctive color patterns of older *Memphis* caterpillars. For example, *M. proserpina, M. aulica,* and *M. beatrix* possess the pattern that we show here. This convergence in morphology exists even though neither their food plants, Monimiaceae, Euphorbiaceae, and Piperaceae, respectively, nor their adult morphology suggests they are closely related. The inference from this statement is that these colors are not a result of heritage (phylogenetic inertia) but instead a result of a common, selective force creating what is perhaps a mimicry complex.

ADULT VOUCHER: 99-SRNP-626; JCM
CATERPILLAR VOUCHER: 02-SRNP-9630; DHJ

41. *MECHANITIS POLYMNIA* – NYMPHALIDAE

Phil DeVries's *The Butterflies of Costa Rica* (1987) has an image of this signature butterfly, *Mechanitis polymnia,* laying a clutch of white eggs on a solanaceous leaf. This species is among the most frequently encountered ithomiine nymphalids fluttering along ACG rain-forest trails and now and then along dry-forest trails. Its forewings, patterned yellow and black, and rusty-orange hindwings represent a coloration repeated in many tens of Batesian and Mullerian mimics of many phylogenetic origins in ACG rain forest and the rest of the wet lowland Neotropics.

This butterfly is the one ithomiine that brings this basic color pattern regularly into ACG dry forest, where it matches well with the occasional visitor *Heliconius hecale* (Nymphalidae). The degree of black patterning, and whether the spots are yellow or rust or orange, makes this a variable model/mimic, in some ways making a bridge between the "tiger-striped" models/mimics and the pattern shown by *H. hecale.* The species are easily distinguished when spread and organized in the museum drawer. The same is not true when they are together in the field and in flight. The taxonomically diverse array of black and orange and yellow butterflies seem to blend into one another.

The eggs of *M. polymnia* are white to pale yellow, occur in clusters, and are very obvious when placed on the upper side of the leaf. Here they may signal, "Do not eat," an aposematic message probably backed up by defensive chemicals sequestered from the caterpillar's solanaceous food plants and passed through the pupa to the adult, and hence on to the eggs. The casual and ostentatious display of the slow-flying, fluttering adult, and its very visible perching on the upper side of leaves also signals, "Do not eat," as evidenced by a pygmy owl that had never seen a butterfly and instinctively rejected immobilized *M. polymnia* adults at a glance (Janzen and Pond 1976). The same owl was offered other species of butterflies in this experiment and responded similarly to a variety of aposematic patterns (see the species account for *Greta morgane,* #95).

The newly hatched caterpillars feed side by side throughout their lives if left unmolested, appearing to be a disorganized tiny herd of gray millipedes. In ACG dry forest, the pale-leafed *Solanum schlechtendalianum,* found on the forest edge, supports the great bulk of the caterpillars of *M. polymnia,* but in ACG rain forest they feed on another five to ten species as well. Given the many tens of species of Solanaceae in ACG dry forest and rain forest, it is evident that *M. polymnia* is, in fact, very choosy about which species of Solanaceae to select for oviposition. Whether the caterpillars should be viewed as some kind of millipede mimic, as suggested by the lateral protuberances, or simply be viewed as odd aposematic caterpillars remains to be examined. Other ithomiines tend to have more normal caterpillars (see *Greta morgane*, #95).

ADULT VOUCHER: 81-SRNP-670; JCM
CATERPILLAR VOUCHER: 04-SRNP-22141; DHJ

42. *CONSUL FABIUS* – NYMPHALIDAE

The adaptive nature of colors and patterns on the wings of butterflies varies greatly from species to species. It also varies within a species. For instance, in certain species the underside (venter) of the wings provides protection through crypsis, namely looking like a leaf. Whereas the leaf mimicry generally conforms to a similar pattern, the upper side (dorsum) of the wings may be very different. *Consul fabius* is an example of what happens when natural selection pushes a nymphalid that mimics a leaf on its underside (also look at *Memphis proserpina*, #40 and *Anaea aidea*, #26) into being a Batesian mimic of a tiger-striped toxic ithomiine or heliconiine nymphalid on its upper side. The ventral wing surface of *C. fabius* resembles a dead, dry leaf clinging to the foliage as if stuck there by a piece of spider web. When it flies, not only does the tiger-stripe pattern of an aposematic butterfly (or one of its mimics) emerge from hiding, but the butterfly itself adopts the leisurely, flopping, sailing flight so characteristic of these models. However, when pursued with a butterfly net, and probably if pursued by a bird, it abruptly drops the visual disguise and launches into the fast wingbeats so characteristic of a pursued *Memphis* or other strong-flying charaxine.

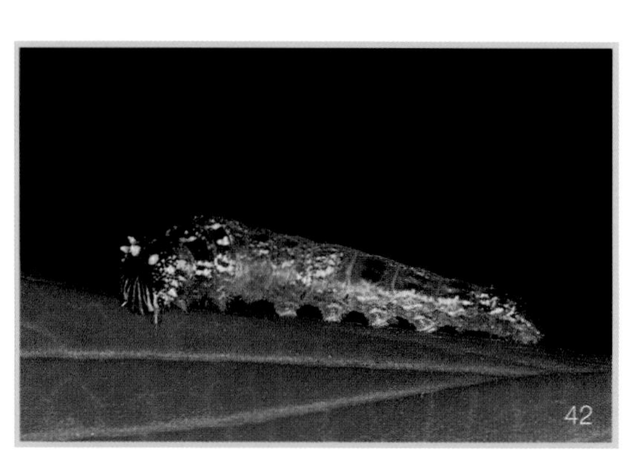

Eggs of *C. fabius* are laid singly on the tips of leaves of many species of *Piper* (Piperaceae) throughout all but the upper cloud forests of the ACG. The caterpillars in their first three instars perch on the tip of the midrib of the leaf they are defoliating, resembling trash stuck there by a bit of spider web and demonstrating the same behavior as other charaxines. Although adult *C. fabius* and its relatives are rarely seen flitting about the forest, the early instars are common and easily located by searching for the characteristic defoliated leaf tip. Shouldn't birds have learned to use the same cues? Perhaps the energetic reward from such a small morsel (a first instar may weigh three to four milligrams) is too trivial.

The later instar *C. fabius* caterpillar behaves much like *Memphis* and other close relatives. It rolls a leaf somewhat like a cigarette, spins a dense blanket of silk, and hides inside this tube during the day. Its massive, multicolored head with prominent tubercules blocks the entrance. Venturing out at night to feed, it does not eat its own nest (as do the much smaller thyridid caterpillars in their tighter and smaller cone-like rolled leaves). The caterpillar we show has been coaxed from its leaf nest. Living in a rolled leaf may be very good protection from birds and other visual predators, but *C. fabius* is certainly not free from parasitoids, which may oviposit directly through the leaf into the caterpillar or come in the front entrance and glue eggs to the face or anterior part of the body, an act performed by Tachinidae.

ADULT VOUCHER: 01-SRNP-15407; JCM
CATERPILLAR VOUCHER: 04-SRNP-42547; JCM

43. *ARSENURA DRUCEI* – SATURNIIDAE

How many ways can evolution create the appearance of a dead and rotting leaf? Easily 10,000, if not a lot more. The wings of moths are a prime example. Natural selection has pushed wing form and color into a myriad of functions, but being either invisible or ignored by vertebrate predators searching by day is always advantageous. Thus, the process of natural selection results in many a moth and butterfly looking like a dead or rotting leaf. Dead and rotting leaves are everywhere and offer no reward to the searching vertebrate. The searching monkey or bird could grab every dead leaf in its vicinity, and perhaps one in 10,000 tries would yield a fat-bodied moth, but it would expend more energy than it would consume.

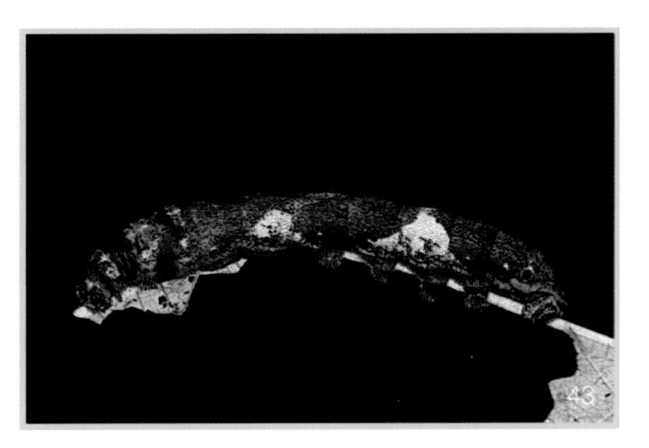

The massive *Arsenura drucei* adult is derived from an equally massive and cryptic caterpillar. The seventeen-centimeter-long caterpillar perches all day among the mosses, lichens, tree bark, and litter of dead leaves in its dank, cold, wet, mushy cloud-forest habitat. Here it is most likely to be overlooked by a bird or monkey. At night the caterpillar meanders up into the canopy to feed on leaves, as does the congener *A. armida* (Costa et al. 2004). The functional nature of the camouflage is maintained in a world that, despite the large number of moths that come to a light trap in this foggy forest, has a low density of vertebrate predators compared to the lowlands below. Why are there fewer predators here? There are far fewer insects per square meter in the cloud forest than at lower elevations, be it dry forest (where there is the most food) or rain forest (where there is less food per day, but food resources are more evenly distributed throughout the year). The fact that there are far fewer insects in the cloud forest, perhaps hardly 1 percent of the biomass of the lowlands, means fewer prey resources and far less predator biomass at the top of the food chain. As *A. drucei* moved away from its lowland tropical arsenurine, or even pre-arsenurine, ancestors (Lemaire 1980) and up into the cold, foggy mountaintops, it not only moved into a world that favored a dark, moist, fungus-ridden look, but it also adapted to the trade-off of nighttime warmth, which it left behind in the lowlands, for a relatively predator-free diurnal sleeping space, which it gained in the highlands.

ADULT VOUCHER: 03-SRNP-21215; JCM
CATERPILLAR VOUCHER: 04-SRNP-32394; JCM

44. *MIMALLO AMILIA* – MIMALLONIDAE

The perching *Mimallo amilia* half wraps its wings around a stem or edge of a leaf tangle and simply disappears into the world of gray-brown, shriveled leaves. What bird or monkey would bother to even look for it? Whoever unwittingly touches this moth will likely recoil in surprise as they feel something soft and silky instead of firm and crispy.

Prior to a human presence, this species was probably a rare animal in the few rocky ridges and cliff faces where a low, sun-loving shrub could survive. Today within the ACG, *M. amilia* is a standard occupant of trashy, abandoned pastures that are dotted with wild guava shrubs, *Psidium guianensis* (Myrtaceae). Look closely at the leaves of *P. guianensis* and find one with the tip partly rolled into a valley with what looks like old spider webbing between the walls of the valley. An even closer look will reveal that this webbing is, in fact, a very intricate close weave of silk and tiny leaf bits, somewhat like the caning of a chair. In some cases, the builder will be hidden within. An early instar constructs this lightly fortified shield against a world replete with predators. It will venture out and forage for leaf parts but zip backward or forward into the shelter when threatened by an approaching figure. As the caterpillar outgrows this lightweight structure, it constructs a strong, thick-walled house of silk, leaf parts, and fecal material. The house is open at both ends (ripped open in our image). The occupant's massive black head blocks one entrance, and a strong rear cuticular plate blocks the other. The black caterpillar is nocturnal. Nonetheless it has orange spiracles, reminiscent of the orange spots along the sides of black pyrrhopygine hesperiid caterpillars like *Jemadia* and *Yanguna,* which likewise live in tough silk and leaf shelters. These caterpillars are only exposed to daylight when the shelter is torn open by a vertebrate looking for prey. It is tempting to include *M. amilia* as a Batesian mimic in the huge ACG mimicry complex of black caterpillars with orange and yellow spots on their sides. The caterpillar is most likely highly edible, hiding in its messy but strong house.

ADULT VOUCHER: 91-SRNP-1751; JCM
CATERPILLAR VOUCHER: 04-SRNP-33603; DHJ

45. *FERENTA CASTULA* – NOCTUIDAE

Ferenta castula is one of those moths that you never know is present until you find the caterpillar. In thousands of light-trap hours in ACG rain forest and lower cloud forest rich in their food-plant vines in the Menispermaceae, *Abuta* and *Disciphania,* an adult has never been collected at a light trap. We are even certain that the parataxonomists working all over Costa Rica have never encountered one because it is such a beautiful and noticeable moth. Only one specimen is present in the huge Neotropical noctuid collection of the United States National Museum in the Smithsonian Institution. We suspect, inferring from other gaudy, large noctuids reared from Menispermaceae in the ACG, which very rarely, if ever, go to lights, that *F. castula* is probably a fruit feeder as an adult. A more successful technique for finding adults would be to deploy baited butterfly traps at night with overripe bananas and mangos or carrion, much as one catches *Archaeoprepona, Prepona, Agrias, Caligo,* and *Memphis,* large diurnal butterflies that do not visit flowers.

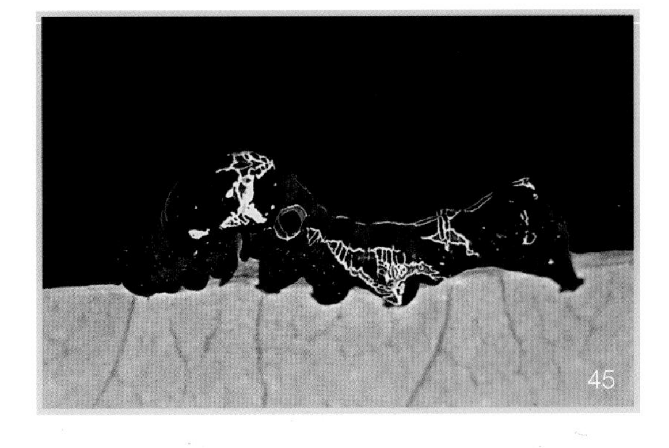

The striking golden-yellow bar on the hindwings of *F. castula* may serve the same function as the yellow (or red or white), strongly marked hindwings of extra-tropical North American fruit-feeding Noctuidae such as *Catocala,* known as "underwing" moths in local parlance. It is widely believed that these moths, when flushed from a diurnal resting position by a foraging bird, present the bright yellow as a flag to be followed, only to disappear abruptly when the escapee alights on a tree trunk and covers the yellow bands with the bark-like pattern of the front wings. But these color-pattern defenses are certainly not haphazardly distributed among ACG moths. For example, they are present in almost all members of the species-rich genus *Gonodonta* (see *Gonodonta pyrgo,* #74). This genus can ascribe to phylogenetic inertia, which means that the ancestor of all *Gonodonta* also had bright yellow patches on the hindwings. But there may be more here than meets the eye. As expected, the more than five species of ACG *Gonodonta* that feed on Menispermaceae have bright yellow patches on their hindwings. It may not be a coincidence that not only does the unrelated *F. castula* have bright yellow hindwings, but so do the much larger noctuids *Eudocima colubra* and *Graphigena gubernatrix,* both of which feed on the same ACG Menispermaceae.

The caterpillar of *F. castula* has evolved a nonnoctuid color pattern, sporting a huge, red false eye on each side of the abdomen, more centrally placed than the false eyespots on many sphingids. The caterpillar neither has the aspect nor the behavior of a snake mimic, as do *Hemeroplanes, Xylophanes, Madoryx, Eumorpha,* and other sphingid snake mimics. Instead, when disturbed it projects its body into an upside-down arch that prominently displays the eye as an object unto itself.

ADULT VOUCHER: 05-SRNP-2373; JCM
CATERPILLAR VOUCHER: 04-SRNP-2444; DHJ

46. *PORPHYROGENES* BURNS01 – HESPERIIDAE

Looking at the female of *Porphyrogenes* BURNS01, you might think it is not a mimic, since there is nothing else that resembles it among the more than 400 species of skippers in the ACG

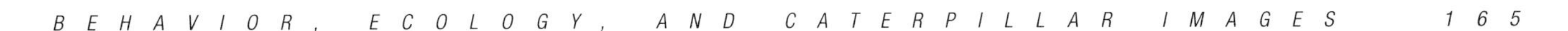

(Janzen et al. 1998, 2005; Burns and Janzen 1999, 2001, 2005a, b; Hebert et al. 2004; Hajibabaei et al. 2006). Think again. Although there is nothing exactly like *P.* BURNS01 pinned and spread, its appearance when buzzing around in the dark understory of the forest or at dusk is another matter. No one has knowingly seen this butterfly in flight in the ACG, despite more than forty having been reared, mostly feeding on *Machaerium seemannii.* There are tens of species of ACG rain-forest hesperiids that are approximately the same: medium-large, dark brown to almost black, and branded with a white band or archipelago of white patches in the middle of the forewing (*Nascus, Celaenorrhinus, Telemiades, Narcosius, Ridens*). Many, if not all, fly at dusk and are denizens of the shaded forest understory. Additionally, they all give the same impression of a dark blur with a white marker, they are fast as lightning, they are very alert to an approaching shape or motion, and they all transmit the message, "I am too fast." Certainly, each has a body large enough to make a meal for an insectivorous bird.

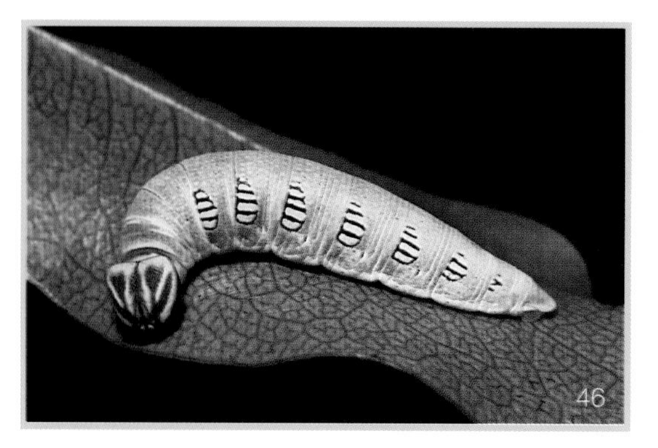

The male of *P.* BURNS01 is quite different from the female. The wings of the male are free of markings, a uniform beige-brown with a slight purplish sheen. As in so many other ACG species of butterflies, the male of *P.* BURNS01 could be seen as a member of a different complex of mimetic hesperiids. The discovery of the *P.* BURNS01 male revealed the tip of a taxonomic iceberg. This species was unknown to science before the ACG caterpillar inventory and is still without an official scientific name.

With the newly discovered male in hand, the obvious next step was to search the medium-large uniformly brown hesperiids in the collection at INBio. This collection was created by setting up light traps throughout Costa Rica. Indeed, specimens of four male *Porphyrogenes* were found. Any butterfly collector would think they looked exactly like *P.* BURNS01, but looks are deceiving. John Burns dissected the genitalia from these four males and found four more species of *Porphyrogenes*—one species per specimen, none previously known from Costa Rica. This, of course, means that somewhere in nature there are four more undiscovered females. Even though these females are unknown, it is a fair guess, based on phylogenetic relatedness, that they will be similar in appearance to the female we show here. According to Bernard Hermier, who studies *Porphyrogenes,* South America contains at least fifteen species of *Porphyrogenes,* and within that mix of species some females (though known from only a very few specimens) are quite similar to those of *P.* BURNS01.

Readers will note that many of the moths and butterflies we portray are undescribed, as is *P.* BURNS01. The taxonomic process is always in motion. As we write and edit this passage in March, 2006, the manuscript presenting the official scientific name for what we call *P.* BURNS01 is in preparation. It will honor Peter Wege of Grand Rapids, Michigan, who has so strongly supported the purchase and conservation of the very forest in which *P.* BURNS01 maintains a healthy breeding population and will continue to do so into perpetuity, thanks to Peter's help.

ADULT VOUCHER: 03-SRNP-6133; JCM
CATERPILLAR VOUCHER: 01-SRNP-1576; DHJ

47. *OXYTENIS MODESTIA* – OXYTENIDAE

Five years into the caterpillar inventory project, we brought home what looked like a small saturniid cocoon made of fine silk and wrapped in a piece of plastic flagging that had been left in a dry-forest successional study plot. We knew all the small saturniids by then and assumed it was just another species of *Hylesia* or *Automeris.* We saved it only to see which parasitoids might emerge. A few days later we had a fine male moth that did not seem quite right for Saturniidae, yet it more or less looked like one. *Oxytenis modestia* and its relatives have been bounced in and out of the Saturniidae (as the subfamily Oxyteninae) for more than a century and today can still be found as Oxytenidae or Oxyteninae in competing modern family-level classifications.

After we found the first dry-forest *Oxytenis* in its cocoon, the caterpillar's biology remained a mystery for two more years. Then as Dan Janzen was walking down a trail in front of Winnie Hallwachs, she said, "Don't you want this one?" He turned, perturbed that he had obviously walked right by something that she then found. His eyes followed hers right to the large *Genipa americana* leaf just under his left elbow, but there was no caterpillar. There was only a grungy bird dropping with a fig seed stuck in the top of it. Then she poked near the fig seed, and the whole grungy mess started to walk away. We had finally found the caterpillar of *O. modestia.* Although the first four instars do well at resembling bird excrement, complete with seeds, the last instar is a marvelous snake mimic, complete with large false eyes.

Both male and female *O. modestia* are extremely variable in color pattern, as are the other two common ACG rain-forest *Oxytenis, O. brepea* and *O. anaemia.* The moths that eclose in the dry season, or in a hot place in the rearing barn in the rainy season, are light beige and yellow, perfect to match the dry and shiny dead leaves both hanging in the tree and lying in the litter (as is the case with *Rothschildia lebeau,* Janzen 1984b). The moths that eclose in the rainy season, or a cool spot in the rearing barn in the dry season, are dark and moldy looking. Collecting them in ignorance from a black light, one is immediately tempted to view the two seasonal forms as different but closely related species.

ADULT VOUCHER: 93-SRNP-7760; JCM
CATERPILLAR VOUCHER: 99-SRNP-11854; DHJ

48. *COPAXA CURVILINEA* – SATURNIIDAE

Put a light trap anywhere in the ACG rain forest below about 1,500 meters elevation and a gray to brown to purplish leaf-like male saturniid is very likely to end up on the sheet, its wings spread almost as though it has been pinned. Such individuals are most commonly *Copaxa rufinans* (if the light is in ACG dry forest, the moth is *Copaxa moinieri*). Indeed, *C. rufinans* is a regular visitor at light traps from tropical Mexican rain forest to South America. Somewhere, and probably not very long ago, a variant of a *C. rufinans*-like ancestor evolved to have a strikingly bent median line running down the front wing and a broad, pale anterior (costal) margin on the forewing's upper side. This is the third species of ACG *Copaxa, C. curvilinea.* This moth is practically identical to *C. rufinans* in color, form, polymorphism, and size, but *C. curvilinea* is much rarer and of a much more restricted distribution, known only from lowland northern Costa Rica (Lemaire 1978). In the ACG the two are fully sympatric in lower-elevation rain forest, but only *C. rufinans* extends up the volcano and inhabits the lower cloud forest.

The caterpillars of *C. curvilinea* are, for all practical purposes, identical to those of *C. rufinans.* Adding to the frustration of identification is the fact that both species feed on the same species of Lauraceae. It is even possible to find several caterpillars of both species on the same individual plant at the same time, feeding side by side. Additional rearing records may show that the relative proportions of each food-plant species used by these two *Copaxa* species may differ.

There are other hints of differences in the biology of these two moths besides their different DNA barcodes. An ominous shift in the abundance and distribution of these two species is occurring. In the late 1990s and early 2000s, nearly all *Copaxa* caterpillars found in the lowlands of the ACG rain forest were those of *C. rufinans.* However, in 2004 and 2005, the great majority have been those of *C. curvilinea.* As climate warming shifts the upper boundary of the lowland rain forest to higher elevations, it may be that each species is moving upslope as the temperatures rise, resulting in *C. curvilinea* becoming proportionately more abundant and reaching further into the zone previously only occupied by *C. rufinans.*

These two species may also support different parasitoid communities. From eighty rearing records, there is only one record of a tachinid fly from *C. curvilinea,* but 480 rearings of *C. rufinans* have yielded a variety of undescribed species of *Blepharipa* and *Hystricia* tachinid flies. Strikingly, neither of the rain-forest species of this pair of *Copaxa* are parasitized by

Hymenoptera, whereas the immediately adjacent *C. moinieri,* which inhabits the dry forest, supports *Enicospilus bozai,* a large and host-specific ichneumonid wasp named in honor of Mario Boza, one of the two founding fathers of Costa Rica's conservation efforts.

ADULT VOUCHER: 02-SRNP-18110; JCM
CATERPILLAR VOUCHER: 00-SRNP-21845; DHJ

49. *MYSCELIA PATTENIA* – NYMPHALIDAE

For many years, this was the mystery butterfly around the house and laboratory in the ACG Administration Area in Sector Santa Rosa's dry forest. Always common, it was present throughout the dry season and much of the rainy season, perched on walls and sucking up water and minerals from the mud at the washing place. All things considered, *Myscelia pattenia* simply had to have a common caterpillar. We did find one pupa once, so we knew its continuous presence meant it probably bred in the ACG dry forest. Then, after seven years, someone found twenty caterpillars on an unidentified bush in the forest understory. Many died, but some were reared to an adult *M. pattenia.* For the moment we had part of the story, but no one could find the bush again.

Another seven years after that, and four kilometers away, we were armed with a better plant taxonomy, and the food plant and life cycle became clear to us. A single understory treelet of *Adelia triloba* (Euphorbiaceae) was found in mid-June, a month after the rainy season began, with many hundreds of caterpillars of *M. pattenia* defoliating it. Known from only seven individuals, *A. triloba* is a very rare ACG dry-forest plant. It appears that the adults of *M. pattenia,* which seem to be everywhere in all seasons except the first months of the rainy season, a time when we now know they exist as caterpillars and pupae, are generated by an extremely concentrated reproductive event using a rare resource. Given the very large number of larvae on a single food-plant treelet, it is likely that the newly mated but old females congregate (via pheromone communication?) at the plant and all lay their eggs at once. Such a reproductive pulse could contribute to local predator satiation. The adults that perch and sail around a water hole or rotting fruit may well be elderly in butterfly terms, as old as however many months have transpired since July. *Myscelia pattenia* appears to be one of the many ACG dry-forest species that is univoltine. The long-lived adults then pass the remainder of the year "hanging out." If we can infer correctly from other species of butterflies that have been more thoroughly studied, it is likely that they do not mate until they have survived the long wait as semi-active adults for the next year's breeding opportunity.

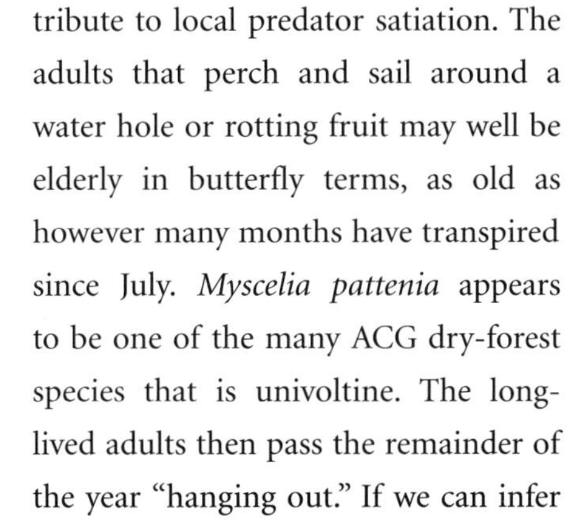

ADULT VOUCHER: 85-SRNP-389; JCM
CATERPILLAR VOUCHER: 01-SRNP-13760; DHJ

50. *HAMADRYAS AMPHINOME* – NYMPHALIDAE

Step out of your car on an ACG road through the forest about anywhere in the lowlands, walk a few hundred meters while watching the large tree trunks, and you have a high chance of bumping into one of the six species of ACG *Hamadryas.* Perched flat against the trunk with wings spread and heads down, their wings, spotted gray, blue, white, and black, match the bark moderately well (hence the generic common name of "calico"). The adults occasionally sortie out in chases and investigations, only to return to a tree trunk to perch, bask, and probe a sap- and ferment-rich tree wound. Some, such as *Hamadryas amphinome,* even have bright flash colors on the underside (red-orange in the case of *H. amphinome*), visible in flight but disappearing abruptly when perched. The ventral markings probably function in the same mode as certain dorsal markings, exhibiting a bright flash of color in flight but hidden when perched by closing the forewings flat over the back to expose the bark-matching pattern.

The loud clicks made by males of *Hamadryas* when they are in flight, including when engaged in aerial combat, have long attracted the attention of butterfly specialists, behaviorists, and ecotourists (Yack et al. 2000). Someone decided to baptize them with the common name of "crackers," probably for the crackling-like noise, even though the common name of "click butterflies" seems a more accurate descriptive name and has been around since the 1950s and probably longer. The mechanism used to create the very audible clicking noise remains unknown, but a tympanum-like structure on the forewing has been identified as the hearing organ (Yack et al. 2000). All the usual suggestions for why male animals produce sound have been applied to *H. amphinome,* in particular territorial display and courtship. None are more than speculation until someone invests serious field and laboratory time in working and living with these "talking" animals.

ADULT VOUCHER: 03-SRNP-13583; JCM
CATERPILLAR VOUCHER: 03-SRNP-13571; DHJ

51. *ZALE PERUNCTA* – NOCTUIDAE

How many ways can Lepidoptera come to look like a piece of tree bark? A striking trait of tree bark is that while it all looks the same to us, there are really as many colors and patterns as there are species of woody plants, multiplied by differences in appearance as bark ages, and then multiplied again by all the ways that dead plant tissue can weather and rot. The answer to our question is that there are seemingly endless numbers of ways to look like tree bark. Since we do not usually eat, use, or otherwise notice it, we have just one word for it—bark. But if you are a bird or monkey that spends its life looking for things to eat that could be sitting on bark, bark is extremely diverse in form, shape, pattern, hue, color, texture, and all those other descriptions we use for clothing—our own bark, so to speak. In the world of butterflies and moths, this means that wings can come from many different evolutionary backgrounds and be many forms, shapes, patterns, hues, colors, and textures and still look like bark. So unless you know every species of tree and what its bark ought to look like, then any one of the thousands of bark-like moths (typically patterned to look like bark on the upper side of the wing) and butterflies (typically patterned to look like bark on the underside of the wing) in the ACG actually do look like bark. Bark is not an edible item to a bird, nor are the wings of a moth or butterfly for that matter, although they do tend to be attached to a potentially edible morsel hidden underneath. An interesting experiment would be to measure how long an insectivorous bird would remain interested, if at all, in bark-patterned wings detached from the moth body versus a less cryptically patterned but not aposematic wing. Our prediction is that the bark-patterned wings would be virtually ignored.

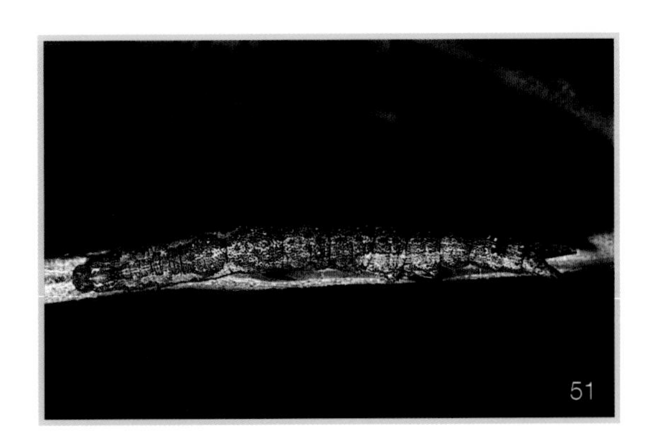

Zale peruncta may be viewed as the epitome, an equal among thousands of equals, of bark likeness. Perched motionless all day against some bark or bark-like plant part, it is left unexplored by insectivores. As a general rule, moths and other insects with such cryptic coloration are highly edible (tasting like raw shrimp if one should care to try, as certainly our ancestors did), and their only real defense until noticed is to avoid being noticed. Once detected by a predator, *Z. peruncta* has only the ability to launch quickly into flight to elude a predator that will not or cannot launch in pursuit. Crypsis, quickness to react, and blinding fast speed are not a guarantee of safety. Along with other insects using this defensive suite of survival tactics, *Z. peruncta* may fall victim to a multi-species foraging flock of birds as they flush prey from the forest floor and foliage. One bird may flush the moth from its resting place, and a nearby bird may snatch it.

ADULT VOUCHER: 04-SRNP-49783; JCM
CATERPILLAR VOUCHER: 01-SRNP-11823; DHJ

52. *LETIS MYCERINA* – NOCTUIDAE

Letis mycerina is a bait moth. Put out a butterfly trap baited with fermenting banana mush (a bit of spit and beer adds some punch to the mixture) and check the trap after dark, when all the butterflies are away sleeping off a day of boozing. Several species of *Letis*, along with other noctuids that do not visit flowers, are likely to be jostling for a place on the bait, their long, stout tongues buried deep in the food, probing for fermentation products (alcohol?), sugars, and other goodies. Both males and females will be there. This is not a search for scarce mineral ions to add to a sperm packet, but rather for energy sustenance for all activities of both males and females. Bait moths in general, and *L. mycerina* in particular, may live for months. Adults of other species may live only a few days to a couple of weeks, dependent on the water and nutrient reserves stored as caterpillars. Bait moths even migrate tens to hundreds of kilometers across Costa Rica following a rainy season generation in the dry forest, which is why they are often collected at light traps placed high in mountain passes between the Pacific and Atlantic sides of Costa Rica.

Looking like a polished slice of banded agate, *L. mycerina* is yet one more example of a moth whose wing resembles tree bark. But you would never associate the female we show with the nearly black male unless you had reared it from a series of caterpillars. It is quite a puzzle as to what selects for great sexual dimorphism in the process of evolution of a large bait moth. Many other species of *Letis* also express polymorphic sexes that are equally difficult to associate with one another. Caterpillars representing three species of *Letis* have been found in the ACG so far. Each species appears to be food-plant specific. One species occupies the new foliage in the tops of rain-forest *Vochysia guatemalensis* trees (Vochysiaceae), a plant not even closely related to the *Inga* food plants of *L. mycerina.* The bark-patterned caterpillars of these large noctuids perch in the daytime on tree trunks and branches, or they will walk all the way down into the ground litter, only venturing out onto the bright green canopy under cover of darkness.

ADULT VOUCHER: 92-SRNP-1750; JCM
CATERPILLAR VOUCHER: 92-SRNP-3051; DHJ

52

53. *HYPERCOMPE ICASIA* – ARCTIIDAE

It seems as though everyone knows the North American woolly bear caterpillar, *Hypercompe scribonia* (known in previous literature as *Ecpantheria scribonia*). It even has its own Wikipedia encyclopedia page. The furry, diurnally active, black and orange caterpillars are so big and ostentatious that they are noticed caterpillaring across an Interstate Highway from a car traveling at 70 miles per hour. It is no more than a myth that they forecast the coming winter weather by the width of the orange band in the middle. The same myth and the same rejection of the myth applies to *Pyrrharctia isabella* (Miller 1995; Miller and Hammond 2000, 2003). *Hypercompe icasia* is the ACG dry-forest version of the North American woolly bear, though it is pure black and was never mythologized as a predictor of the weather.

Diurnally and nocturnally active, one wonders if they are always on the go. Females lay many hundreds of tiny eggs in large patches just about anywhere. The tiny first instars nibble on just about any species of plant, or so it seems. Most must die early in life, probably because they are appealing morsels to predators. They grow extremely slowly. The conspicuous older caterpillars march from plant to plant, eating some foliage here and eating some other foliage over there. It is very likely that by being able to eat about any species of plant, they are getting almost no nutrients from the plant, or that their digestive process is severely impacted by the undetoxified defensive chemicals, or both. By the time that a few have survived long enough to be noticed by us, approaching the diameter of a pencil and about three centimeters long, most other species in the habitat have long since passed through the caterpillar stage, or even advanced into a second generation during the first half of the rainy season. At this larger size, if confined to the particular species of plant on which it was encountered, *H. icasia* seems to be capable of feeding and successfully pupating while feasting on a single species of plant.

Extreme polyphagy is a trait characteristic of less than 1 percent of the species of ACG Lepidoptera (Janzen 2003). Polyphagous feeders are not just rampant, random, opportunistic primary consumers, however. *Hypercompe icasia* shares the ACG dry forest with the smaller, gray, and multicolored *Hypercompe albescens,* whose caterpillar feeds in the same seemingly polyphagous manner and walks a great deal. Over the decades the inventory project has accumulated a list of hundreds of species of plants in many tens of families whose foliage is eaten by these two extremely polyphagous species. Yet, when the food plant lists for each species are compared, there is almost no overlap between them. In other words, they are both extremely polyphagous in their food-plant use, but each species appears to be specialized as a polyphagous feeder on some set of plants chosen on the basis of some traits not known to us.

The caterpillar of *H. icasia* seems to be protected from vertebrate predators by its hairy black exterior, a Batesian mimic of an urticating caterpillar. The adult is dressed in classic aposematic coloration, becomes catatonic ("plays possum"), and if provoked, "bleeds" a yellow, bitter fluid from its joints, a phenomenon know as reflex bleeding. It could very well be that the toxic chemicals in the blood are manufactured by the moth. The moth seems to possess nasty chemicals irrespective of whether the food plant was one rich in defensive chemistry (e.g., Solanaceae) or quite lacking in it (e.g., Sterculiaceae). This topic is ripe for experimental investigation since it may be that *H. icasia* could easily be viewed as a Batesian mimic for its harmless spines and as a Mullerian mimic for its body chemistry.

ADULT VOUCHER: 98-SRNP-2530; JCM
CATERPILLAR VOUCHER: 04-SRNP-35295; DHJ

54. *NEONERITA DORSIPUNCTA* – ARCTIIDAE

There is a human phenomenon associated with conducting a species inventory project for a large, complex, and diverse place such as the ACG. The inventory has to start somewhere temporally, taxonomically, and geographically. In our case, many years of that geographical somewhere was the Sector Santa Rosa dry forest. The staff of the inventory project documented and became intimately familiar with many species. These species then became dry-forest species in our minds. As the inventory project expanded to cover other parts of the ACG, some of these species were then met in rain forest, cloud forest, or various kinds of intergrades between the three major ecosystems. In these "new" places, the species then seemed out of place, although often they were not. It has become second nature to say, "What a surprise to find that dry-forest species in the rain forest." *Neonerita dorsipuncta* is such a species. Although it is a denizen of both dry forest and rain forest, it is almost impossible for us to avoid thinking of it as a dry-forest species.

54

Encountered repeatedly in Sector Santa Rosa as a gregarious caterpillar feeding on the evergreen leaves of sapling *Sideroxylon capiri* (previously known as *Mastichodendron capiri*, or *tempisque* in local Spanish), the bright red and black caterpillars align themselves side by side while wolfing down the leaf, making a very visible display. Presumably each individual does better facing a world of carnivores (some not seeing so well) as part of a visually enlarged display than it would by wandering about by itself ostentatiously in full daylight. On this point, however, it does seem that if they are separated from their sibs, the individual *N. dorsipuncta* caterpillars feed and grow very readily. In other words, their social behavior as siblings does not appear to be a necessity to overcome a tough leaf or plant chemistry, for example.

The natural history of *N. dorsipuncta* in Santa Rosa dry forest illustrates the geo-temporal sequence of discovery. There are only three species of sapotaceous trees in the upland ACG dry forest: *S. capiri*, *Manilkara chicle*, and *Chrysophyllum brenesi*. Of these, *N. dorsipuncta* females lay eggs almost exclusively on *S. capiri*. This led to our thinking of *N. dorsipuncta* as functionally monophagous. We thought it wouldn't eat foliage of a species outside this family of plants. However, as the inventory project entered the ACG rain forest, we found *N. dorsipuncta* caterpillars eating not only other species of Sapotaceae but also two species of Myrtaceae. We must admit that the myrtaceous foodplant records represent only two observations and might well have been caterpillars that strayed from their natal food plant. Nonetheless, the species is not monophagous. Here again, new information leads to revised perceptions and a better understanding of the species at hand.

ADULT VOUCHER: 99-SRNP-5624; JCM
CATERPILLAR VOUCHER: 03-SRNP-3875; DHJ

55. *PROTOGRAPHIUM PHILOLAUS* – PAPILIONIDAE

Anywhere that pawpaw, the annonaceous treelet *Asimina triloba*, grows in the eastern United States, the observant butterfly watcher knows it is eaten by the caterpillar of the black and white long-tailed swallowtail, *Protographium marcellus*. Its Neotropical analogue and perhaps its closest living relative, *P. philolaus* (also previously known as *Eurytides philolaus* and *Neographium philolaus*) is an occasional denizen of ACG dry forest.

The ACG coincides with the southern extent of its range. The northern extent of its range is northern Mexico. Throughout this range the species inhabits lowland dry tropical forest and thus possesses a remarkable physiological robustness for relatively extreme climatic fluctuation. This robustness may explain how a species like *P. marcellus,* with a tropical ancestral lineage, manages to exist far into the central United States. It is unclear if the ability to tolerate the harsh, hot, and dry part of the year as a dormant adult began with range expansion to the north by a rain-forest *Protographium* that subsequently returned to the most environmentally fluctuating part of the tropics, the dry forest, or whether by adapting to tropical dry forest, it was preadapted for a later invasion, along with its tropical food plant *A. triloba,* of the north. Pawpaw is the only member of the large and very old tropical family Annonaceae to have invaded cold northern climates.

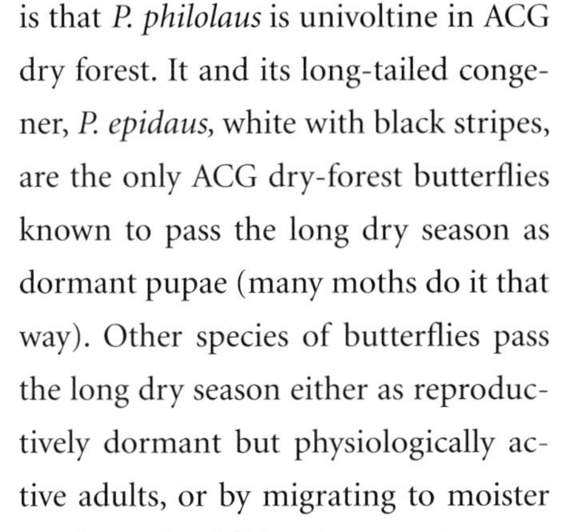

Adult *P. philolaus* are on the wing in the ACG from about a month before the rains begin (April) until a few weeks after the rains have begun (early June), and then gone. But where do they go? The eggs are deposited on the broad, densely tomentose leaves of *Sapranthus palanga* and *Annona purpurea* (both Annonaceae), with a few on *Annona holosericea.* They hatch in a few days. The caterpillar requires about three weeks to develop. The caterpillars do not feed on the other common annonaceous treelet in the same habitat, the glabrous- and narrow-leafed *Annona reticulata.* The tough, hard pupa is found tied with silk to a dead leaf in the litter rather than on an exposed leaf or twig still connected to the plant, as is normal for most butterflies. In nature, the pupa then remains dormant in the litter all the way through the remainder of the rainy season and the first three-quarters of the long dry season. In the rearing barns, however, the pupae remain dormant for a month to six months. It is not clear what cue causes certain adults to eclose early. What is clear is that *P. philolaus* is univoltine in ACG dry forest. It and its long-tailed congener, *P. epidaus,* white with black stripes, are the only ACG dry-forest butterflies known to pass the long dry season as dormant pupae (many moths do it that way). Other species of butterflies pass the long dry season either as reproductively dormant but physiologically active adults, or by migrating to moister ecosystems where they pass through additional generations, or, in a very few species of Hesperiidae, as dormant prepupae.

Another life-history trait is shared by *P. philolaus* and *P. epidaus.* They are parasitized by an undescribed species of *Trogus* (Ichneumonidae) that attacks them exclusively. Not surprisingly, the wasp larva remains dormant in the dormant butterfly. Adult wasps emerge about the same time as the adult butterfly.

ADULT VOUCHER: 84-SRNP-384; JCM
CATERPILLAR VOUCHER: 85-SRNP-278; DHJ

56. *PROTOGRAPHIUM MARCHANDI* – PAPILIONIDAE

This gorgeous, long-tailed swallowtail occasionally adorns the muddy riverbanks of ACG rain forest. Males can be found puddling, sucking ion-rich water from the mud to include sodium in the sperm packets delivered during mating. Following mating, many of the sperm packets are dissolved by the female and the constituent components are used as nutrients for egg development (see our account of *Phoebis philea,* #82). The few Costa Rican museum specimens of *Protographium marchandi* are males that were collected while they were puddling. The staff of the inventory project despaired of ever finding the caterpillars, despite the butterfly's range from Mexico to South America. This despair was intensified because we had so thoroughly searched the foliage of species in the Annonaceae and Lauraceae, their presumed and most likely food plants in ACG rain forest. These plants are fed on by caterpillars of the three other known ACG *Protographium: P. philolaus* (#55), *P. epidaus,* and *P. marcellus.*

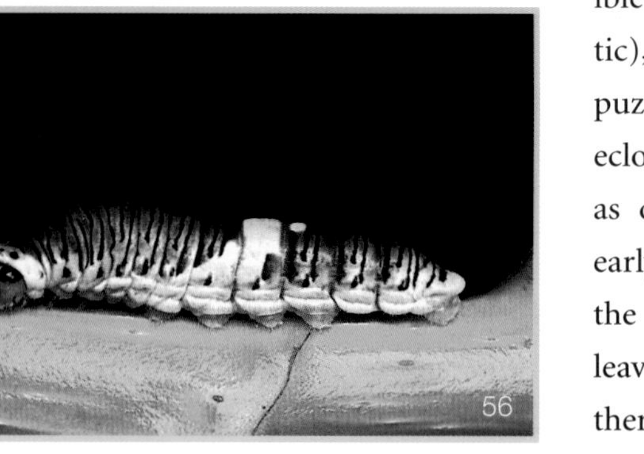

Finally, in 2004, despair was replaced by euphoria when a strange, new papilionid caterpillar was found feeding on the bright green leaves of *Talauma gloriensis,* a large and scarce rain-forest tree in the Magnoliaceae. At the time of collection, we did not know the identity of this caterpillar, but we did speculate that it might be our missing swallowtail. Two weeks after pupation, we had the answer in the form of an adult. It was *P. marchandi.* Oddly, this caterpillar restricted itself to the very youngest leaves of *Talauma.* Even stranger, *T. gloriensis* crowns typically shed their leaves and grow a new leaf crop in one short burst during February and March, so food for this rain-forest caterpillar is available for just two months but also for only one generation per year. This explains why the adult males are seen on the riverbanks only during the first part of the rain-forest dry season. They emerge from pupae that have been dormant for ten to eleven months, having survived all the predator and fungal challenges that the rain forest can throw at a highly edible and motionless (though very cryptic), naked pupa silked to a twig. It is puzzling that some of the adults do eclose within a few weeks of pupation, as did the first one we reared. This early eclosion happens when, at least in the ACG, there are almost no new leaves on *T. gloriensis.* Furthermore, there has been no sign of *P. marchandi* caterpillars on the few new leaves that appear later in the season. Perhaps the adults from nondiapause pupae migrate to some other rain forest that is seasonally timed to take advantage of new magnolia foliage and establish another generation. The female specimen shown here was reared and is among the first ever to be seen. It is likely that the normal life of a female adult is played out high in the canopy, with only the occasional male descending to a muddy riverbank.

ADULT VOUCHER: 05-SRNP-1165; JCM
CATERPILLAR VOUCHER: 05-SRNP-1159; DHJ

57. *KLONEUS BABAYAGA* – SPHINGIDAE

Twenty years ago, Jean-Marie Cadiou, the European coauthor of the master checklist of Sphingidae of the world (Kitching and Cadiou 2000), visited Philadelphia to see what sphingids had been encountered at the beginning of the ACG biodiversity inventory project. At that time, the primary focus of the project was on the dry forest. His first question was, "Have you found a big brown sphingid?" We had not. Why the interest in this moth? Only two had ever been collected in Central America over the last century, one in Nicaragua and one in Costa Rica. From a Central American standpoint, this rare South American species would be a real prize. The big brown moth of interest was *Kloneus babayaga,* a species that has been bounced around among subfamilies and left unsettled taxonomically. Finding the caterpillar would provide valuable information and, in itself, mark a grand achievement. The challenge was on.

The general aspect of adult *K. babayaga* did suggest that it might be related to *Pachylia* and therefore feed on *Ficus, Brosimum, Castilla,* or some other big-leaf Moraceae, as do the three species of ACG *Pachylia.* We had no idea if it even occurred in the ACG. But then, in 2004, a moth was dumped out of an envelope from the *gusanero's* freezer at the rain-forest Estación San Gerardo, and a fine male *K. babayaga* was staring us in the face. It took only a click on the computer to check the database and see that Gloria Sihezar Araya had described the field-collected early instar as "*Pachylia ficus,*" eating *Ficus pertusa* in the rain forest. We knew then that *K. babayaga* was a close relative of *Pachylia.* The adult also resembles *Pachylia* but with strongly serrate wing margins. Gloria went back to that specific tree and searched it thoroughly. Another caterpillar was found. Results from DNA barcoding of the two specimens also showed that the species is very similar to *P. ficus.*

In a teasing manner, we joked to Jean-Marie Cadiou that perhaps *K. babayaga* should be sunk into *Pachylia.* This statement elicited a classic taxonomic tome of a letter on 11 August 2004, "As to systematics, I strongly object to the prestigious *Kloneus* being downgraded to the status of a mere *Pachylia* . . . Besides which the phrase '*Kloneus babayaga*' was coined by Skinner to mean 'son of a bitch,' in reference to Preston Clark, and it would be a pity to lose that piece of entomological history."

Despite additional intense searching and the rearing of everything looking remotely like a *Pachylia ficus* caterpillar, no more specimens of *K. babayaga* have been seen in the past two years. But now we know it is there. We also know that it does not come to light traps, which undoubtedly contributes to why it is so rarely seen.

ADULT VOUCHER: 04-SRNP-2375; JCM
CATERPILLAR VOUCHER: 04-SRNP-60316; JCM

58. *MADORYX PLUTONIUS* – SPHINGIDAE

The gorgeous adult of *Madoryx plutonius* has, as with the other two species of *Madoryx, M. bubastus* and *M. oiclus,* evolved to resemble a dead, blackened leaf. The silver spot so obvious on the dorsal surface of the forewing looks like a hole in a leaf, letting through bright sunlight, when the moth perches in the field. This identical appearance has evolved independently in *Calledema plusia* (#20).

From a total of seven caterpillars found over a six-year period, we thought we had this species figured out. Each caterpillar was found feeding on ACG rain-forest Melastomataceae and perching cryptically on the beige twigs. We watched each of these caterpillars spin its fine silk cocoon against the tree trunk, quite atypical for a sphingid. It turns out that the other two species of ACG *Madoryx* share this trait for cocoon construction and placement. Then one day Petrona Ríos Castro found a sphingid egg glued neatly to the newly expanding leaves of *Vochysia ferruginea* (Vochysiaceae). A sphingid egg on *Vochysia?* What sphingid caterpillar could possibly feed in the crown of this tall rain-forest tree? We all watched the little *gusano* grow into a huge *Madoryx* caterpillar that when disturbed revealed a huge false eyespot on each side of the thorax but when not disturbed resembled a thick twig. Then the caterpillar died, leaving us without a name and wondering if the egg placement had been a simple error of oviposition. Perhaps this food plant was really not the right food plant—species of *Vochysia* are not related to any of the other *Madoryx* food plants and lack foliage containing toxic chemicals, a trait exhibited by typical sphingid food plants. The other species of *Madoryx* used food plants also known to lack toxic chemicals, however. Perhaps what *Vochysia* has in common with Verbenaceae, Melastomataceae, and tree Bignoniaceae is that it is relatively bland.

58

Two years later, just to see what he could see, Osvaldo Espinoza Obando climbed a *Vochysia guatemalensis* tree. He saw six caterpillars fervently munching on new and old foliage. In one sighting he not only confirmed a food plant but also nearly doubled the total number of field-collected *M. plutonius.* So once again we have a single species feeding on the foliage of two quite different rain-forest plant families. We now have a clue as to why *Madoryx* caterpillars spin a strong cocoon. They live and stay up in the canopy among the big branches of a big tree, rather than tempting fate as a pupa in the predator-rich litter below. Maybe species of *Madoryx* are simply tree-inhabiting sphingids, putting them ecologically far from the many more typical ACG sphingids that feed on herbs, vines, shrubs, saplings, and treelets. As yet we know little of the biology of these rare rain-forest sphingids.

ADULT VOUCHER: 03-SRNP-11544; JCM
CATERPILLAR VOUCHER: 03-SRNP-21707; DHJ

59. *OXYNETRA HOPFFERI* – HESPERIIDAE

If there ever was a skipper butterfly adult that does not look like a skipper butterfly adult, it is the male of *Oxynetra hopfferi.* The first ACG specimen to be seen was walking up the inside of its rearing bag, which was hanging on a clothesline in the Estación Cacao rearing barn. We immediately thought the voucher numbers were mixed up, and that someone had put an arctiid caterpillar into a bag that was supposed to have a new species of pyrrhopygine skipper in it. It was colored like a brilliant red, black, and white diurnal ctenuchine arctiid, it was the right size for an arctiid, and it walked with exactly the same gait of ostentatious display, holding its wings out and back nearly horizontally, just as do aposematic arctiids.

59

The caterpillar of *O. hopfferi* with its brilliant yellow rings lives high in the cloud forest on Volcán Cacao and probably in other cloud forests in Costa Rica, where it eats the mature foliage of *Prunus annularis* (Rosaceae). Although *P. annularis* is rich in cyanogenic glycosides, the foliage smelling of the cyanide of crushed almonds, everything else about this skipper butterfly says that it is a true Batesian mimic of diurnal arctiid moths. There is no arctiid that it specifically mimics, a situation common among mimics of diurnal arctiids. Its walking behavior is also an exacting mimicry of diurnal Arctiidae. It is likely, however, that tests will show it to be edible, differing from the diurnal arctiids, which sequester the cyanogenic glycosides in their food and are presumably inedible.

We portray the female of *O. hopfferi,* totally black except for one narrow, red ring on the abdomen and white palpi. She appears to be a Batesian mimic of yet a different group of arctiids, the black wasp mimics *Macrocneme, Antichloris,* and *Poliopastea.* Again, like the male, she walks calmly over the foliage and rearing bag with the same stature as these ctenuchine arctiids. Extreme sexual dimorphism within a species of Batesian mimic is common in tropical nature. However, it should be noted that the basic black color of this cloud-forest skipper that lives in a cold habitat at high elevations is also common among other diurnal Lepidoptera (see *Creonpyge creon,* #60) and a variety of other insects in the same ecosystem, and may well be functional in the process of gathering heat through insolation. Knowing which came first in evolutionary time, the ostentatious black or the heat-absorbing black, will probably remain lost to the ages.

ADULT VOUCHER: 02-SRNP-23286; JCM
CATERPILLAR VOUCHER: 02-SRNP-23109; DHJ

60. *CREONPYGE CREON* – HESPERIIDAE

The fog rolls through the dense trees and small-leaved evergreen shrubs like water. The temperature is cold and dropping. Not a bird stirs, there is no sun in sight, and the air on the top of

Volcán Cacao is thin. As the wind pushes and howls at intervals, a butterfly drops out of the fog, perches on a leaf, and turns to scrutinize you. Droplets of water bead on the wings and speckle the body. It slowly turns and aims the flat surface of its hairy body and spread wings at the location of the invisible sun, presumably soaking up radiant energy in a world that seems to have none. This is not butterfly weather, but it is home to the deep black *Creonpyge creon,* with a blue sheen and a big red dot glowing on the rear part of the hindwing. Living in this environment, *C. creon* must be the most cold-hardy of all ACG diurnal butterflies. It is not a night-flying moth or crepuscular butterfly venturing into the daylight hours. Though never common, it is the one butterfly to be found active in bad weather in the uppermost cloud forest on the top of Volcán Cacao, at an elevation of 1,400–1,500 meters. Neither adults nor their caterpillars have been seen at lower elevations. Sadly, it will probably be one of the first ACG butterflies to go extinct locally as the warm lowlands march inexorably up the slopes of the volcano, a result of the measurable effects of global climate change (Pounds et al. 2006).

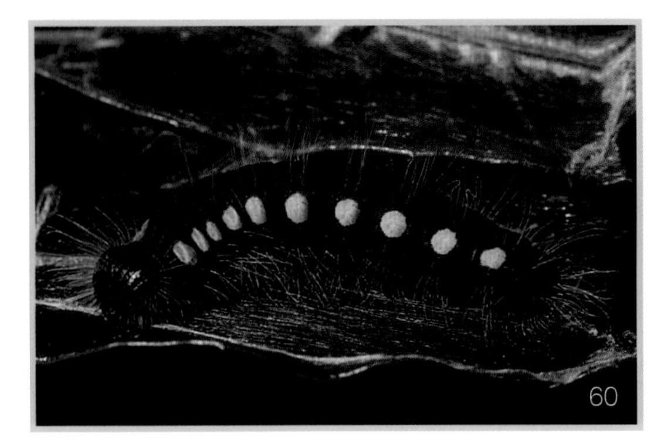

The caterpillar of *C. creon* was the pyrrhopygine caterpillar that until recently had escaped the inventory project. We knew it was there, having seen the adults. Precisely when we were summarizing and preparing the photographic plates for the ACG pyrrhopygine Hesperiidae (Burns and Janzen 2001), Harry Ramirez found a gorgeous black-bodied, laterally orange-spotted pyrrhopygine caterpillar feeding on the leaves of *Dendropanax querceti,* an araceous shrub at the top of Volcán Cacao. We guessed it was *C. creon* and published the image, but left it without a name. We later discovered that we could just as easily have been wrong, since the caterpillar could have belonged to the other volcano-top pyrrhopygine, *Oxynetra hopfferi,* at that time unknown in the ACG. The caterpillar of *C. creon* is unambiguously a member of a caterpillar mimicry complex found throughout the ACG—black-bodied with lateral yellow and orange spots. This caterpillar differs only in its very slow growth rate, which is to be expected when living at largely refrigerator-like temperatures.

The tops of mountains are ecological islands, and the insects (and most other animals and all plants) that occupy them are effectively insular populations. It is likely that when the *C. creon* population that we guess is on the top of Volcán Rincón de la Vieja (or its complex) is DNA barcoded and compared with that on Volcán Cacao, the two may well turn out to be different. The taxonomist will then be confronted with the conundrum of whether they are different enough to be viewed as two species.

ADULT VOUCHER: 03-SRNP-3109; JCM
CATERPILLAR VOUCHER: 01-SRNP-7142; DHJ

61. *BUNGALOTIS ASTYLOS* – HESPERIIDAE

This big nocturnal (and perhaps crepuscular) hesperiid is actually a common species, but you would never know it from the very small number of specimens in collections. Nearly all of the museum specimens were obtained by collecting the occasional individual that showed up at a light trap along with thousands of moths. The female of *Bungalotis astylos* is very different from the male that is shown in our portrait. The female is dark chocolate brown with a broken band of whitish-hyaline areas angling across the forewing, somewhat as in *Astraptes* INGCUP (#9). She also has rounded wings, in contrast to the narrowed forewings and the elongate, backward-projecting hindwings of the male.

The puzzle lies in understanding the selective forces that drove the nocturnal male and female of *B. astylos* into such different colors and aspects. Extraordinarily sexually dimorphic Hesperiidae are dotted across the ACG butterfly landscape but are unusual, even though it is often possible to determine the gender of a hesperiid at a glance by the small differences in wing shape, aspect, and color. However, many of the sexually dimorphic species are as *B. astylos*, a brown male and a darker female with the semblance of a white band angling across the forewing. Since two other species of ACG *Bungalotis, B. midas* and *B. quadratum,* are equally sexually dimorphic, it is reasonable to postulate that the original driving force may have been lost in the ecology of the ancient event when these three species first split apart.

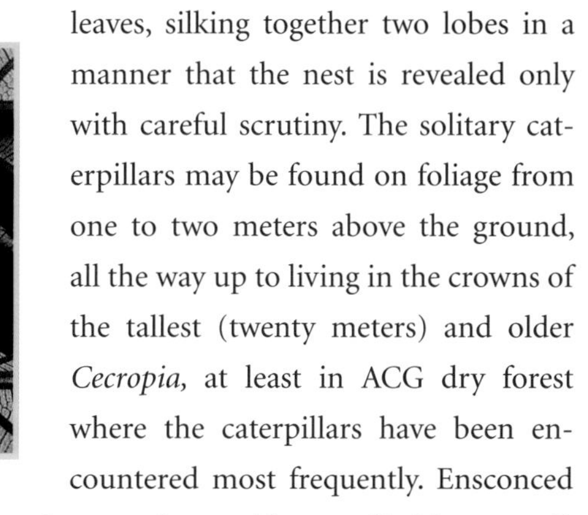

The mottled, reticulated, cream and greenish-white caterpillar of *B. astylos,* with its strikingly yellow-striped head, appears to be specialized to feed on the mature leaves of the three lowland species of ACG *Cecropia,* an ant plant in dry forest and rain forest. None of the other species of Cecropiaceae are used. Tolerating, or tolerated by, the vicious *Azteca* ants that keep *Cecropia* clean of vines and nonspecialist herbivores (Janzen 1969), *B. astylos* tends to live in a fold of the palmately lobed leaves, silking together two lobes in a manner that the nest is revealed only with careful scrutiny. The solitary caterpillars may be found on foliage from one to two meters above the ground, all the way up to living in the crowns of the tallest (twenty meters) and older *Cecropia,* at least in ACG dry forest where the caterpillars have been encountered most frequently. Ensconced in their leafy houses that are thoroughly patrolled by ants, *B. astylos* caterpillars are notably free of parasitoids, having none that are specialists and attacked by almost no generalists.

ADULT VOUCHER: 01-SRNP-1878; JCM
CATERPILLAR VOUCHER: 97-SRNP-6183; DHJ

62. *BUNGALOTIS DIOPHORUS* – HESPERIIDAE

We have never seen an adult *Bungalotis diophorus* in the wild, perched or flying, in the ACG. Surely they are there. Their biological tracks, otherwise known as eggs, larvae, and pupae, are common in the ACG rain forest. The immatures are found only on saplings and treelets of *Simarouba amara* (Simarubaceae), though they might on occasion use the dry-forest *Simarouba glauca* where it abuts against the *S. amara* population. The leaf nests of an early instar are visible on the *S. amara* leaflets, which are neatly arranged along the long rachis like the barbs of a feather. This is because the underside of the leaflet is a very light green, and contrasts with the dark green upper side. When the caterpillar folds over a cut-out portion of the leaflet, the exposed underside reveals the presence of the nest. Later instars silk one leaflet lightly down on top of another, rendering the nest almost invisible unless one concentrates on noticing that the leaflets are not in such an orderly array as usual. The clues revealing the presence of a caterpillar may or may not be used by foraging birds, but they certainly help the *gusanero*.

Why do we not see the very different-appearing male, which strongly resembles the male of *Bungalotis astylos*, and female *B. diophorus* adults? We suspect that they fly in the very late afternoon when it is nearly dark, or even at night. The suggestion of nocturnal activity is less likely, because to our knowledge none have been collected at lights. The female, with her bright white spots, is probably very visible to the monochromatically red-brown male, even in the low light of evening. If the wing is viewed at a steep angle, many but not all of the white spots glow like a roadside reflector.

Here we identify this butterfly as *B. diophorus*, but we suspect that it really is an undescribed species. The problem is that it looks very much like *B. diophorus* found in Brazil, a specimen of which is the holotype for the species name. It looks so much like it that almost anyone would identify the ACG species as that. However, the skipper taxonomist for the inventory project, John M. Burns at the Smithsonian Institution, has found slight differences between the ACG specimens and the Brazilian specimens, so we await a new name in the coming years. This example underlines a chronic problem with Central American butterflies. Many closely resemble the type specimen that was collected in South America, and only very close taxonomic scrutiny, and often substantial further collecting, can determine if the Central American population is a separate species.

ADULT VOUCHER: 02-SRNP-15042; JCM
CATERPILLAR VOUCHER: 01-SRNP-2435; DHJ

63. *SARSINA PURPURASCENS* – LYMANTRIIDAE

Everyone knows the gypsy moth caterpillar, *Lymantria dispar,* a European introduction into North America that periodically defoliates large expanses of forest that some of us don't want to see defoliated. *Sarsina purpurascens* is an ACG cousin to the gypsy moth, but it occurs at excruciatingly low density. The caterpillar inventory project has found about twenty caterpillars in twenty-five years. The species occurs in both dry forest and rain forest and appears to be a food-plant specialist, having been found feeding on only four species of Myrtaceae in the genera *Eugenia, Psidium,* and *Calyptranthes.* By contrast, *L. dispar* caterpillars, especially in their penultimate and ultimate instars, appear to be able to eat an extremely wide selection of extratropical foliage (Miller and Hanson 1989). If the gypsy moth happened to occur in ACG dry forest, it would be able to eat the foliage of only a small fraction of the species, and even fewer in ACG rain forest. No, we are not going to try that experiment. Actually, in the context of ecological analogs, the experiment is in place. A species of ACG dry-forest lymantriid, *Orgyia* JANZEN01, has a biology very similar to that of *L. dispar,* complete with localized population explosions that defoliate *Quercus oleoides* (Fagaceae) and *Byrsonima crassifolia* (Malpighiaceae) but spreads no further, even though individual caterpillars can and do eat a few other species of plants.

Just as *S. purpurascens* adults make it through the day by looking like a beige, dead leaf half-wrapped around a twig, the densely hairy caterpillars sporting a brown, gray, and black tweed pattern likewise are extremely cryptic. The head-to-tail hairs on the caterpillar project forward and down over the head, fully hiding it, and project laterally down to the twig covering the feet. Clinging tightly to a medium-diameter twig or tree bark, the caterpillar resembles an inedible swelling of dead plant, something very abundant in all of the ACG ecosystems.

The eggs are laid in patches of forty or so, side by side in a very orderly grid. Emerging simultaneously eight days later, the pale green and hairy first instars feed side by side for a day and then go their separate ways. The early instars are leaf green but transform to shaggy brown-gray-black-tweed at the molt from penultimate to ultimate instar. Whereas such hairy, cryptic caterpillars are sprinkled across the ACG caterpillar topography, the *Sarsina* pupa is not. The pupa is a mosaic of light beige and nearly black, and is suspended by a few silk strands on the surface of a green leaf. It appears to be a pupa that would be cryptic in litter, but what is it doing on the surface of a green leaf? When seen by the insectivore (using Janzen's field experiences with students in classes on tropical ecology as surrogates for the foraging bird or monkey), it is ignored or remains undetected because it has the appearance of an eclosed butterfly pupa, an empty and nutrient-free shell dangling among a few strands of silk or fungus.

ADULT VOUCHER: 04-SRNP-47815; JCM
CATERPILLAR VOUCHER: 97-SRNP-5858; DHJ

64. *REJECTARIA ATRAX* – NOCTUIDAE

As its dark wings with a single line resembling fungal hyphae imply, *Rejectaria atrax* is a moth of the deeply shaded understory of ACG rain forest. Perching on the ground and water-soaked litter as much as on dark tree trunks, it is one of those blurry, black things that your footsteps flush, only to quickly disappear by sitting again a few meters away, essentially invisible. Occasionally you may meet an individual with its proboscis pushed deep into a fallen rotten fruit on the forest floor. Although they are typically nocturnal moths, that behavior is overwhelmed by the opportunity to sip sugar and alcohol-laden juices. The male we show has huge palpi extended back up over its head, a signature trait of this group of noctuids, the Herminiinae.

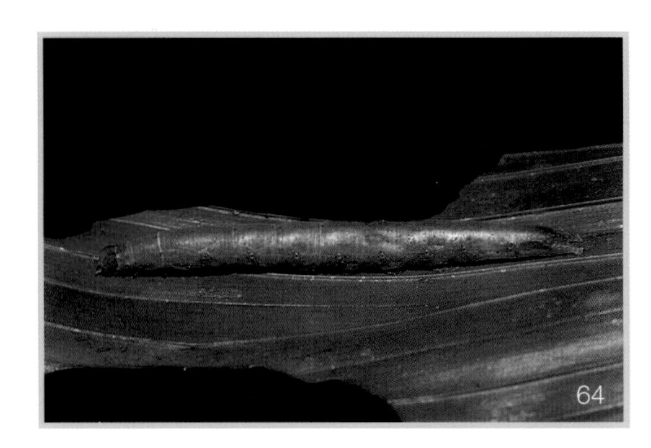

What is this group of noctuids? Primitive plants have a mixed reputation, perhaps somewhat deserved, of not being fed on by many species of herbivorous insects. Patches of ferns, mosses, or *Selaginella* are not places that one automatically searches when looking for caterpillars. However, the practice of the ACG caterpillar inventory project is to search just about everywhere rather than where things are supposed to be. Thus, the *gusaneros* have frequently encountered various species of *Rejectaria* and other hermiines feeding on ferns. There are even species that thrive on a diet of rotten log (probably laced with fungal hyphae) or a diet of rotten leaves in the litter. The caterpillar of *R. atrax* is a rather ordinary green worm, but unlike most ordinary green worms, it feeds on the leaves of three species of Cyclanthaceae, a very ancient and primitive monocot. It shares its proclivity for Cyclanthaceae with two congeners, *R. villosa* and *R.* JANZEN01, as well as with five species of very distinctive Hesperiidae and not much else.

The long, leaf-green caterpillars of *R. atrax* are an excellent example of the sort of green, nondescript caterpillars that ACG small birds bring to their nests, in contrast to the large numbers of very brightly colored caterpillars that human observers (and presumably foraging vertebrate predators) encounter. Leaf-green caterpillars are particularly difficult to find in the daytime. They are often hidden on the underside of leaves or in folds of leaves. They are much easier to find at night, when the waxy cuticle reflects the light of a flashlight in a quite different manner than does the cuticle of a green leaf. In the last days of larval development, during the prepupal phase of the last instar, the green larva turns bright pink-brown, presumably to be camouflaged against the leaf litter it may have to navigate for the few hours it wanders about prior to settling into a place to initiate pupation.

ADULT VOUCHER: 05-SRNP-35567; JCM
CATERPILLAR VOUCHER: 03-SRNP-11914; DHJ

65. *ERBESSA SALVINI* – NOTODONTIDAE

Erbessa salvini bears a species epithet honoring Salvin of the famous pair of collectors and taxonomists Godman and Salvin, who found so much and told us so much about the biodiversity of the Neotropics in the late 1800s. Their work lives on in the Biologia Centrali Americana (go to http://www.sil.si.edu). Compare *E. salvini* with *Zunacetha annulata* (#66). Instinct says that in no way are they related, deviating from the classical notodontid color pattern and gestalt (see *Bardaxima perses,* #19 and *Calledema plusia,* #20) in the usual dramatic way of dioptine notodontids. But these two day-flying moths of the same subfamily have evolved into different mimicry complexes, perhaps by different predators in different places, or simply by the first color mutations taking one in one direction and the other in another direction, all in the context of what by chance was present to mimic, or according to what by chance was present to drive the mimicry. All three species of ACG *Erbessa* exhibit the same basic pattern of violent yellow patches on coal black, and flutter ostentatiously along path edges, serving as an easy target for a predator if edible and an ignored object if inedible. However, once again, the presumed inedibility could use some supporting evidence from detailed field tests. The sole melastomataceous food plant of the caterpillars, *Henriettea tuberculosa,* is not rich in defensive chemicals that are toxic to vertebrates. If *E. salvini* is toxic, the adult or caterpillar is probably ginning up its own nasty molecules.

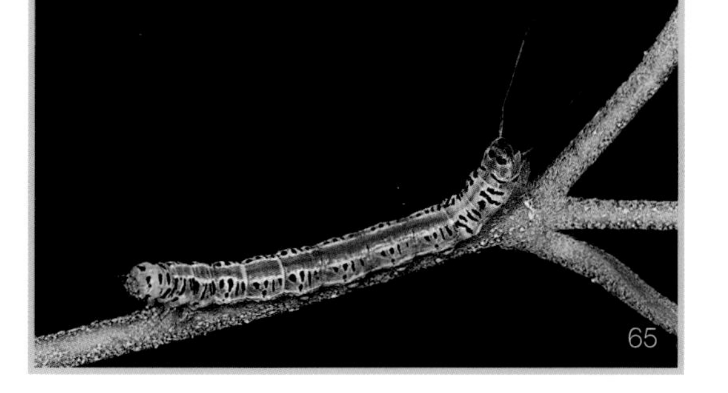

Whereas the adult of *E. salvini* has gone the direction of one kind of aposematism or mimicry (see Miller 1996), the caterpillar has gone another. First, with its orange head and rear end, and green, pale-striped and variegated body, it conforms to the appearance of classic dioptine caterpillars, which form part of a huge mimicry ring of ACG caterpillars. Whether any or all of these caterpillars are Batesian mimics or Mullerian models/mimics, or simply avoided based on predator instinct, is yet to be discovered.

Many notodontid caterpillars have terminal prolegs modified into other structures, creating a signature profile for the family. The most posterior (anal) prolegs of *E. salvini* are enormously extended into what can best be described as white-tipped, long, black antennae, making the rear of the caterpillar appear to be the head end of some large insect. But what is the value of having the rear look like the head? If the predator strikes the rear, more rubbery and less laden with nerve receptors, the caterpillar has a higher chance of wiggling or falling away into the tangle of vegetation below, into which many a bird would be reluctant to dive in pursuit. However, the white tip may signal something more significant. The caterpillar of *Hylesia continua,* a common, highly urticating, and often ostentatious black and white caterpillar in the same habitat, has the anterior black and white-tipped scoli enormously extended and clearly wiggles them about as part of its aposematic display. The rear of *E. salvini* may be a copy and locked into the same defense.

ADULT VOUCHER: 04-SRNP-55738; JCM
CATERPILLAR VOUCHER: 04-SRNP-55959; JCM

66. *ZUNACETHA ANNULATA* – NOTODONTIDAE

Adult Notodontidae are infamously considered dull in appearance, meaning that they resemble dead foliage or tree bark to escape detection by diurnal predators. There is one exceptional subfamily, the dioptine notodontids (Miller 1996). Occurring only in ACG rain forest and lower cloud forest, all the species in this subfamily appear to be mimics of aposematic Lepidoptera or are aposematic themselves (the relevant feeding tests have never been conducted). They spend much of the day flying, courting, displaying, or otherwise being visible. The combination of the characteristics of bright colors coupled with the behavior of never appearing in light traps strongly suggest that they do not have much of a night life, if any at all, though surprises do remain everywhere in the tropics. *Zunacetha annulata* is a bright white dioptine notodontid with a bit of yellow. If simply aposematic, the species is good at it. If a mimic, the model is no longer there. During population outbreaks, an adult seems to be sitting on every bush. The local birds do not appear to be making a living by grabbing them, as easy as it would be.

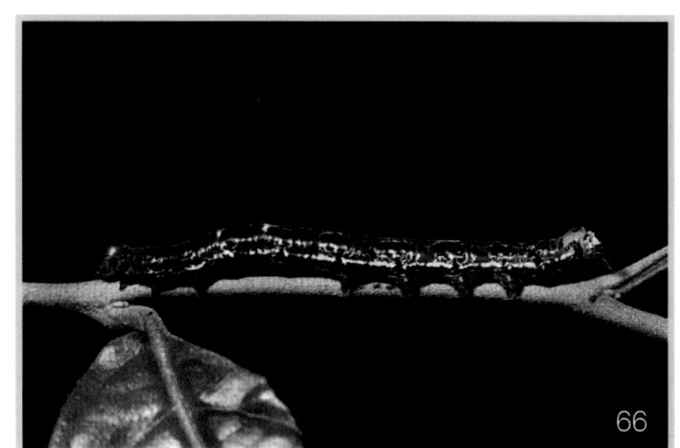

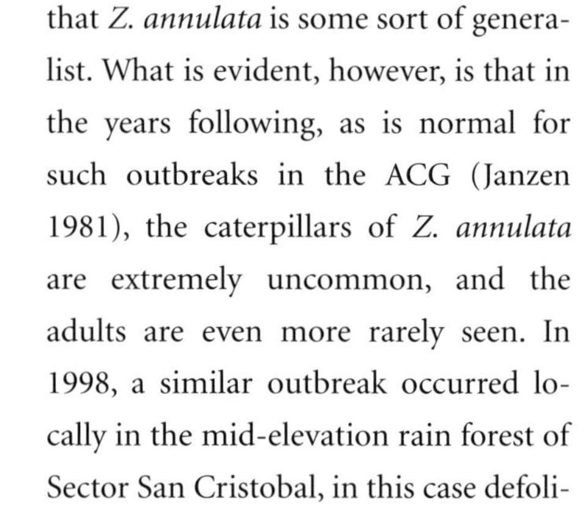

The same applies to the caterpillars during their outbreaks (Wolda and Foster 1978). The ovipositional food plants of *Z. annulata* are several species of understory *Hybanthus,* shrublets in the Violaceae. In June of 1988 there were handfuls of *Z. annulata* caterpillars on every plant of *H. hespericlivus,* a common species in the lower cloud forest and mid-elevation rain forest on the west side of Volcán Cacao and of Volcán Orosí. The caterpillars defoliated every individual plant, and then the starving caterpillars walked off in all directions, sampling as they went. They found many species of plants on which they could eat and grow to pupation, leading to the false impression that *Z. annulata* is some sort of generalist. What is evident, however, is that in the years following, as is normal for such outbreaks in the ACG (Janzen 1981), the caterpillars of *Z. annulata* are extremely uncommon, and the adults are even more rarely seen. In 1998, a similar outbreak occurred locally in the mid-elevation rain forest of Sector San Cristobal, in this case defoliating *H. denticulatus.* Since birds do not feast on the outbreak caterpillars, and since 244 rearings have produced only one parasitoid, it seems evident that neither carnivores of adults nor parasitoids of caterpillars are responsible for their low density between outbreaks. At the end of an outbreak, there is no *Hybanthus* foliage left for the next generation. It is tempting to suspect that the *Hybanthus* response to a defoliation event might be to produce foliage too toxic to eat for years to come, but surely this effect must dissipate over the years. What blocks yet another outbreak is a total mystery.

ADULT VOUCHER: 98-SRNP-2296.05; JCM
CATERPILLAR VOUCHER: 04-SRNP-35602; DHJ

67. *TELEMIADES CHRYSORRHOEA* – HESPERIIDAE

An adult *Telemiades chrysorrhoea,* the butterfly with the unpronounceable species epithet (to make it easier just drop the "ho"), when viewed against the backdrop of all skippers, falls easily into the morph of "dark butterfly with yellow rear," owned by many tens of tropical Hesperiidae (Burns and Janzen 2005b). To what degree this pattern is part of predator avoidance, the mating dance, or thermoregulation remains to be teased out, but whichever it is, the pattern is widespread throughout the hot lowland tropics, and *T. chrysorrhoea* exhibits a much blacker aspect than most.

On the cold upper parts of Volcán Cacao, butterflies turn black, so to speak. A black body absorbs radiant energy from the sun quickly, a fine trick in a world that most closely resembles the inside of a refrigerator. While all lowland species of *Telemiades* are brown with scattered hyaline spots, the same shades of brown that decorate hundreds of species of medium-sized Hesperiidae, *T. chrysorrhoea,* a denizen of the volcano top, is so dark brown that it is essentially a black butterfly. The back edge of the hindwings, top and bottom, shows a flashy yellow area, in short, a pattern resulting from a sort of optimization of selective forces involving the physical environment, predator regime, and courtship rituals.

Telemiades chrysorrhoea has a look-alike, *Telemiades gallius,* that has been found only twice in the ACG. Those sightings were at intermediate elevations. The adults are so similar in appearance that *T. chrysorrhoea* was, for a short time, considered to be *T. gallius.* But *T. gallius* bears three minute white hyaline dots near the tip of the forewing, probably something that the opposite sexes are well aware of. DNA barcoding has confirmed that they are different species.

The taxonomic status of a species is under constant scrutiny. When the ACG specimens of *T. chrysorrhoea* were barcoded, the DNA sequences fell into two groups instead of the expected one. How could this animal possibly be two species? It is not. Researchers are currently addressing how it is that the PCR process reads the true DNA barcode of the male mitochondria, but reads the slightly different sequence from the analogous pseudogene in the female nuclear genome. This does not result in confusion of *T. chrysorrhoea* with other species, but it does give the false impression that there are two entities within one.

ADULT VOUCHER: 03-SRNP-4126; JCM
CATERPILLAR VOUCHER: 95-SRNP-350; DHJ

68. *PROTEIDES MERCURIUS* – HESPERIIDAE

This big skipper in the *Epargyreus-Drephalys-Aguna-Chioides-Proteides* color pattern complex, resembling one of the eight species of ACG *Epargyreus,* mystified us for the first sixteen years of the ACG caterpillar inventory project. Seen here and there visiting flowers and feeding on bird droppings at all times of the year, the large *Proteides mercurius* adults just had to have a large, legume-eating caterpillar somewhere in the ACG dry forest. But we simply did not bump into it through the usual haphazard search. All the other big skipper caterpillars had long since been collected and reared. Then one day a *gusanero* felt something plump inside a dark brown, dead leaflet rolled and hanging among the green leaves on a branch of *Dalbergia.* That something was a *P. mercurius* caterpillar, many individuals of which had been hiding in front of our very eyes for more than a decade, eating the leaves of *Andira, Dalbergia, Piscidia,* and nine species of *Lonchocarpus* trees.

68

The large caterpillar is dark green, dark olive, or nearly black. The head is black, with dark red false eyes, and the abdomen sports a thin red lateral line. These dark colors, instead of the greens and leaf-yellows of its close relatives, match the background and cave-like microhabitat of the inside of a dark brown to black dead leaf. When the roll is torn, instead of fleeing, the caterpillar remains motionless, looking like more torn, dead leaf, though a close look down the entrance presents the viewer with a pair of glaring red eyes. Even the pupa blends well with rotting leaves. Covered by a dense white powder, it looks like a moldy, rotten pupa or a dead insect. Again, a close look down the front of the roll presents a fierce, white face with two big, deep black eyes looking right at you.

The underside of the wings of *P. mercurius* show striking white markings, as do many of the other look-alike species. They are probably part of the distinctive pattern that tells flycatchers and other small insectivorous birds that it is much too fast to bother chasing. This diurnal butterfly can be observed openly sitting on a leaf as it sucks on a bird dropping that has just been softened by its own defecation, or sipping nectar from flowers, or just skippering along through the low vegetation looking for mates and oviposition sites.

ADULT VOUCHER: 02-SRNP-4516; JCM
CATERPILLAR VOUCHER: 04-SRNP-2133; DHJ

69. *ZARETIS ELLOPS* – NYMPHALIDAE

One of the most masterful disguise artists and the supreme "dead-leaf" butterfly of ACG dry forest is *Zaretis ellops.* Appearing to be some sort of slightly off-color orange-yellow pierid as she flutters among the foliage searching for an oviposition site, *Z. ellops* lands, presses its wings together, and now appears to be a dead leaf. The monochrome orange-yellow upper side disappears when the wings are held vertically together, and the wings display the dead-leaf pattern, reticulate brown on yellow-beige. Even the tails on the hindwing appear to be the leaf petiole.

The disguise is not just one look, but two, each adapted to the appropriate season. The dry forest's color spectrum varies at different times of year. During the long dry season, when many trees are leafless and full sun penetrates much of the understory, the world is yellow-beige with a smattering of other bright, light colors. There are no *Z. ellops* caterpillars that look like dark twigs, and no bright green pupae at this time, presumably owing to the lack of green in the habitat and the largely leafless nature of their food plants, primarily *Casearia nitida* and *C. arguta* (previously Flacourtiaceae, presently Salicaceae). The population consists of reproductively inactive adults, feeding a bit on spoiling fruits and at sap-rich fermentation sites on broken tree branches and trunks. The underside of the wings match the pattern of light yellow, dead leaves. Some physiological cue at the end of the rainy season leads to this light coloration, as with the short-lived saturniid moths that eclose at the end of the long dry season (Janzen 1984b). Then come the rains about 15 May. The few surviving *Z. ellops* mate and oviposit on the newly flushing foliage and then die. Their offspring, eclosing in the dark world of the full rainy season, with its dense shade and dark mold colors, are not bright yellow but instead a darker orange-brown on the underside, still looking like a dead leaf. As amazing as it sounds, *Z. ellops* is no loner in this trick. Seasonally based polymorphisms in multibrooded species have been known for centuries (Shapiro 1976, Ruszczyk et al. 2004).

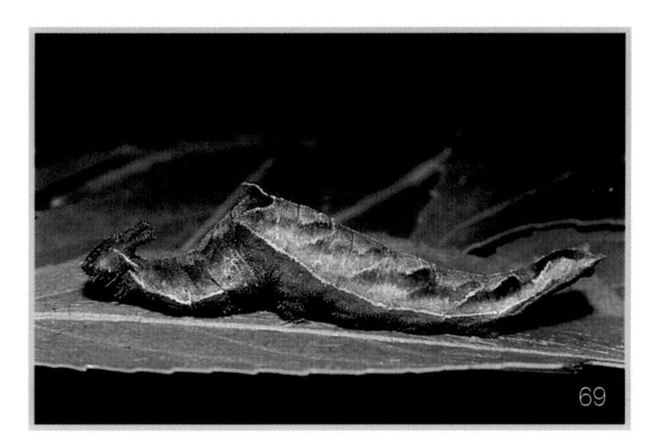

ADULT VOUCHER: 91-SRNP-883; JCM
CATERPILLAR VOUCHER: 00-SRNP-15442; DHJ

70. *ZARETIS ITYS* – NYMPHALIDAE

If the entire ACG had been just dry forest, the *Zaretis* story would end with *Zaretis ellops* (#69). But the ACG contains rain forest and cloud forest as well, and evolutionary patterns between congeneric species emerge as *Zaretis* is followed into ACG rain forest. The classic view (DeVries 1987) is that ACG dry forest is occupied by *Z. ellops* and ACG rain forest is occupied by *Z. itys.* Comparing the undersides of *Z. ellops* and *Z. itys* females reveals an oft-repeated story of dry-forest versus rain-forest adaptations. The female of *Z. itys* is also a marvelous dead-leaf mimic, but she is a mimic of a dark, moldy, rain-forest leaf, complete with markings emulating a combination of insect and fungal damage. However, unlike *Z. ellops,* as the dry and wet periods come and go in ACG rain forest, *Z. itys* does not produce a second morph. The singularity of its morph may be because the dry and wet aspects of the seasons in the rain forest are not as extreme as they are in the dry forest.

Another species of dead-leaf variant, essentially a *Z. ellops* look-alike in the ACG rain forest, has long annoyed butterfly taxonomists. Is it just a *Z. ellops* morph under the continuous influence of rainy-season environmental conditions, so that it looks like *Z. ellops* in the ACG dry forest in the rainy season (see *Z. ellops*, #69)? In contrast to *Z. itys*, it has an underside very similar to that of the *Z. ellops* we show, though it is nearly as darkly marked as the darkest *Z. ellops* from ACG dry forest. This look-alike morph has been called *Z. isidora*, when relegated to full species status (but, for example in DeVries 1987, it was placed under *Z. ellops*). DNA barcodes show that *Z. isidora* reared in the ACG rain forest are not only very different from *Z. ellops*, but also that *Z. isidora* itself is two species, one primarily in the deepest ACG rain forest in Sector San Cristobal and Rincón Rainforest, and the other primarily in the interface zone between dry and rain forest in Sector Pitilla and Del Oro. They currently carry the interim names of *Z.* JANZEN01 and *Z.* JANZEN02 and can be distinguished by the facies of the males. The species that occupies the interface between dry and rain forest, *Z.* JANZEN01, is more responsive to seasonality and produces light-colored morphs that are almost identical in appearance to the rainy-season morphs of *Z. ellops*.

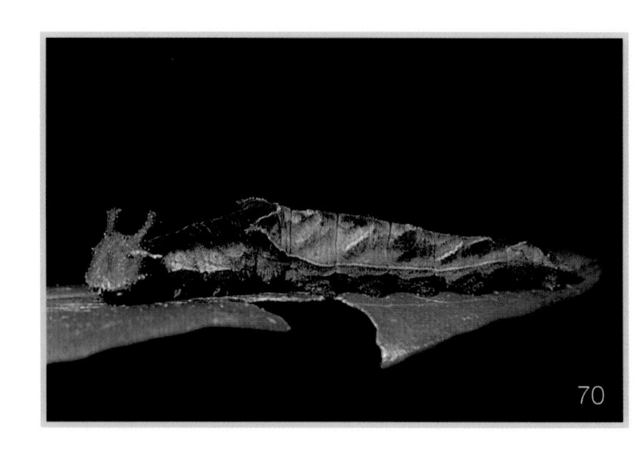

The caterpillar food plants are actually of little assistance in solving the puzzle of identities. All three species of ACG rainforest *Zaretis* feed on *Casearia arborea*, found in the area that wraps around the north side of Volcán Orosí to just barely come in contact with some of the dry-forest species of *Casearia* eaten by caterpillars of *Z. ellops*. Any specimen of what appears to be *Z. ellops* or *Z.* JANZEN01 (*Z. isidora* of old) needs to be barcoded to know its identity with certainty.

ADULT VOUCHER: 03-SRNP-10355; JCM
CATERPILLAR VOUCHER: 03-SRNP-37289; DHJ

71. *ENTHEUS MATHO* – HESPERIIDAE

Cross your eyes a bit, perhaps take off your glasses, and compare this female of *Entheus matho* with the male of *Dismorphia amphiona* (#72), a pierid Batesian mimic of inedible, tiger-striped heliconiines and ithomiines. There is no avoiding the similarity of these two very unrelated butterflies when they are seen in casual flight. It is a mimetic association that generally goes unnoticed by the collector because hesperiids and pierids tend to be the focus of persons with very different interests. Since both species are Batesian mimics, they occur at the very low density expected in comparison with their ostentatious and sometimes seemingly abundant inedible, aposematic models.

The ACG brings the difficulty of such mimetic comparisons yet further into focus. Caterpillars of *E. matho* feed on the very young foliage of saplings of large trees of *Alfaroa guanacastensis,* in the walnut and hickory nut family (Juglandaceae), in ACG cloud forest. The population also extends down into mid-elevation rain forest at Estación Pitilla, where caterpillars feed on *Oreomunnea pterocarpa* (Juglandaceae) and *Eschweilera neei* (Lecythidaceae). The real bulk of the population lives up high and in the cold, where *Dismorphia amphiona* does not, and where tiger-striped models are also only the occasional visitor. So it would appear that we have a mimic that lives in an environment largely devoid of inedible models.

The ACG is a small-scale version of the transcontinental phenomenon of tropical birds migrating to higher latitudes to nest in the comparatively food-rich and predator-poor spring (Janzen 1993). Although the genetic histories and natural histories driving this phenomenon are still blurry, it is clear that birds move around in the ACG as the seasons ply their courses, and among those movements are elevational migrations. The bird that today is studiously deciding not to sally after a male or female *E. matho,* perched or fluttering among the foliage in the cloud-forest understory, may well have been, as recently as a month earlier, searching for insects in the mid-elevation rain-forest understory, which is rich in tiger-striped models and their mimics. Furthermore, the bird may well carry with it, wherever it goes, a genetically hard-wired program that says "no" when that color pattern crosses its mental radar. If it does say no, it is certain that such a pattern was not burned into its DNA by selection occurring in the cloud forest on Volcán Cacao. The color pattern of *E. matho* and the other two very similar species of ACG *Entheus* (*E.* BURNS02 and *E.* BURNS03) have an ecologically and evolutionarily tangled history, one that is not easily clarified.

ADULT VOUCHER: 04-SRNP-32406; JCM
CATERPILLAR VOUCHER: 95-SRNP-349; DHJ

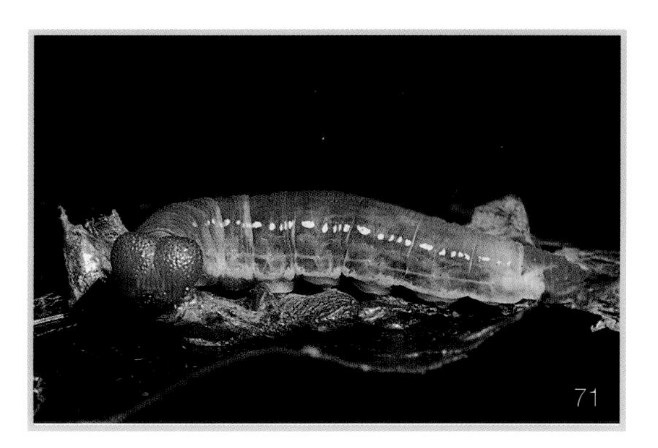

72. *DISMORPHIA AMPHIONA* – PIERIDAE

That butterfly is a pierid? Yes, *Dismorphia amphiona* is a perfectly good pierid, exhibiting the classic pierid look of a soft velvet and uniform green caterpillar feeding on the leaves of *Inga* (Fabaceae), a perfectly good legume, the food plant family for a very large number of the world's Pieridae. However, in the tropics, where it seems that practically every butterfly has evolved to look like some other butterfly, *Dismorphia* is famous for having evolved away from the typical white or yellow, fast-flying, ostentatious pierid morph and into a morph resembling a toxic heliconiine or ithomiine, as has a nymphalid, *Consul fabius* (#42).

It is a fair guess that the ancestors of *Dismorphia* used their white and yellow wings as part of their recognition and courtship rituals, just as typical pierids do today. The male of *D. amphiona* still signals himself to the female, a nearly flawless heliconiine and ithomiine mimic in her own right, by the large white area on the anterior part of the hindwing, an area that he covers with the tiger-striped forewing except when fluttering in a courtship mode near the female. In other words, at most times when living out his life, he is visually a tiger-striped heliconiine or ithomiine, just as she appears to be. But when it comes to courtship, he drops back into the ancestral mode of behavior and flashes colors that she is presumably tuned into.

If you are sitting quietly in the forest, and a *D. amphiona* should happen to flop by, looking like its tiger-striped models, grab your net and take a swing. You are likely to miss. The startled butterfly instantly puts aside its ruse of being a fluttering heliconiine or ithomiine and reverts to the frantic wingbeat so characteristic of white and yellow pierids around the world. When the defense is visual, both sexes are mimics; when the defense is physical, highly powered flight is in order.

ADULT VOUCHER: 99-SRNP-4235; JCM
CATERPILLAR VOUCHER: 05-SRNP-33141; JCM

73. *STROPHOCERUS THERMESIA* – NOTODONTIDAE

One of the many ways we have obtained a scientific name for a moth collected from an ACG light trap or reared from a field-collected caterpillar has been to go to the set of drawers containing that family of moths in the incredible Lepidoptera collection at the United States National Museum within the Smithsonian Institution. There we pull out drawer after drawer until, after hundreds of drawers, there is a match to the moth we are comparing. The matching specimen is side by side with other specimens that are conspecifics, exhibiting whatever range of variation that exist in the species, including sexual dimorphisms and polymorphisms. When going through the Notodontidae this way, aside from the bright dioptines (see *Erbessa salvini,* #65), we see drawer after drawer of dead leaf and bark mimics. Then we pull a drawer and there is a unit tray filled with moths with bright rust-red front wings, *Strophocerus thermesia.* Adjacent will be several other unit trays with other species of *Strophocerus,* all based on the same body plan and wing pattern.

The striking uniform coloration of the *S. thermesia* front wings is not conspicuous in nature. This is, in part, because when the wings are held together, tent-like over the back, the light gray speckled hind margin of the forewings divides the monochrome into two smaller patches of rusty reddish brown. This may be hard to appreciate with older and faded specimens, but a freshly eclosed *S. thermesia* really stands out among typical notodontids. The white and largely pattern-free hindwings, a dull monochrome as characteristic of so many moths that hide all day, may simply be the result of a lack of selection for pattern, since they are invisible when the moth is motionless. Alternatively, the pale hindwings may serve the same function—"now you see me, now you don't"—as the bright yellow markings on the hindwings of some moths (see *Gonodonta pyrgo*, #74).

The caterpillar of *S. thermesia* also stands out among notodontids. A smattering of species of large ACG notodontid caterpillars exhibit false eyespots, and this caterpillar is right in there doing its part. As with the caterpillar of *Xylophanes tersa* (Sphingidae) (go to http://janzen.sas.upenn.edu), the caterpillar of *S. thermesia* has a row of eight false eyes on its lateral line, building on the eye-like beginning offered by the lateral spiracles, one to a segment. The effectiveness of these false eyes as defensive traits is unknown.

ADULT VOUCHER: 81-SRNP-927; JCM
CATERPILLAR VOUCHER: 81-SRNP-927; JCM

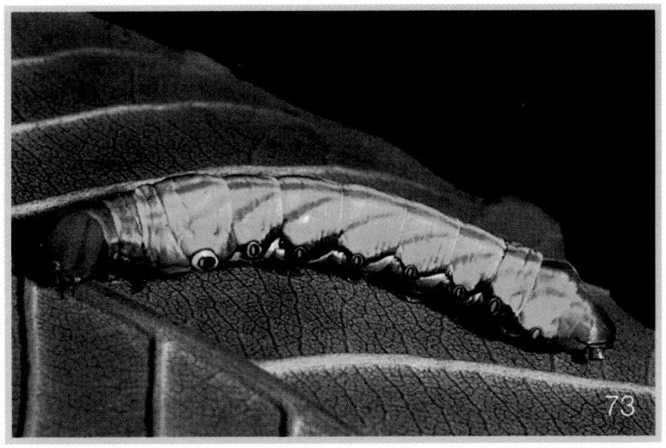

73

74

74. *GONODONTA PYRGO* – NOCTUIDAE

Staring bleary-eyed at the light-trap sheet at an elevation of 2,000 meters along a mountain pass somewhere in Costa Rica at 2 AM, shivering and hearing the condensing fog drip off the leaf tips all around you, you may realize that the medium-sized noctuid perched in front of you known as *Gonodonta pyrgo*, brown forewing with a cream leading edge, certainly did not grow up in this wet and soupy vegetation. The food plants do not grow here. Caterpillars of *G. pyrgo* are common feeders on, and even severe defoliators of, lowland dry-forest Annonaceae, *Annona* and the tropical pawpaws, *Sapranthus*. The caterpillars are most prevalent in the first month of the rainy season and are only occasionally encountered during the remainder of the year. The adults migrate. They come to light traps in the high cold mountain passes. Half are males and half are females (Janzen 1987a,b; 1988a). The adults are emigrants from the food-rich and predator-poor dry-forest lowlands, where a single generation developed. They will fly all night across the high, cold passes and drop down into the rain-forest lowlands to pass through another generation, perhaps two or three, before some of the descendants fly back to the dry forest with the first rains a year later to repeat the cycle. This is why you meet *G. pyrgo* perched on the wall beneath airport lights in San Jose and trapped inside restaurant windows and at gas stations all over the country. Thus, the ACG serves as a net exporter of *G. pyrgo*, importing some unknown smaller number of them ten to eleven months later.

The caterpillar of *G. pyrgo* displays another widespread entomological phenomenon. In certain years, the adults arrive at the dry forest in great abundance, depositing enough eggs to create a very large number of surviving caterpillars that certainly are not escaping from visually orienting predators by being rare and cryptic. The caterpillars perch above and below the leaves, feed day and night, and can occur in the hundreds to tens of thousands in the crown of a large tree. As is commonly the case with such species, the caterpillars come in a great multitude of colors. This polymorphism is presumably selected for through the formation of search images by birds foraging on one particular color pattern, thereby favoring "the others." But once again, as is so often the case with ostentatious caterpillars, it is not really clear if this is simply a polymorphism in a caterpillar that obtains safety in numbers or whether all the bright yellows, reds and blacks, and spots and stripes of *G. pyrgo* caterpillars are simply part of mimicry of a great array of Mullerian and Batesian genetic models, all of which occur abundantly in ACG dry forest. Such a puzzle will never be solved with armchair speculation, but will require extensive detailed observation of predator behavior, coupled with a modicum of inference about the long chain of selective events dotting the evolutionary histories of these species.

ADULT VOUCHER: 05-SRNP-22980; JCM
CATERPILLAR VOUCHER: 83-SRNP-893; DHJ

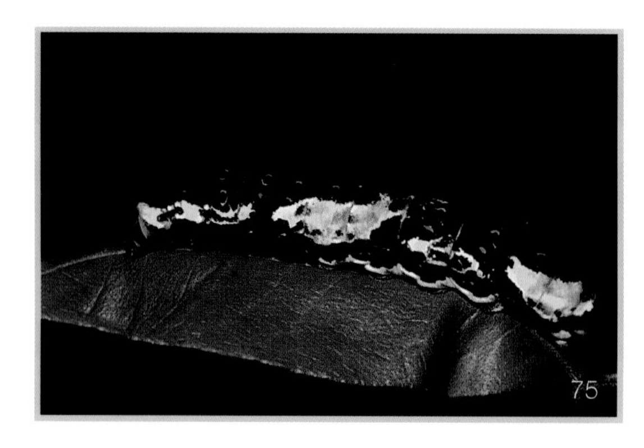

75. *HERACLIDES ASTYALUS* – PAPILIONIDAE

When the ACG was established, one of the conservation-oriented goals was to purchase property. In so doing, numerous small frontier homesteads were added to the ACG. Each of these properties had *Citrus* trees, in particular, limes, oranges, and tangerines. Through the years the presence of these trees has altered the distribution of certain species of butterflies. Introduced from the Old World, these plants have long been associated with humans, making the associated fauna an anthropogenic fauna. The foliage of various citrus trees serves as preferred oviposition site for some of the Neotropical butterflies that are specialists on native rutaceous shrubs and treelets. For example, *Heraclides astyalus* is one of the four species of *Heraclides* (also *H. cresphontes, H. anchisiades,* and *H. androgeus*) that make use of this "new" food. It is entirely possible that one or all four of the *Citrus*-eating *Heraclides* established populations in the ACG by following the orange plantations and backyard *Citrus* trees as they penetrated Costa Rican biospace over the past four centuries. However, we will note that *H. astyalus* and *H. cresphontes* are the most likely candidates for this kind of human-associated spread of a native species. This statement is based on observations that a huge proportion of the current ACG populations of these two species is supported by *Citrus* trees instead of the wild Rutaceae. To date there is no evidence that two other ACG look-alikes, *H. torquatus* and *H. rhodostictus,* have similarly broadened their food-plant tastes.

The female we show is a classic color pattern thought to be

a Batesian mimic of inedible *Aristolochia*-feeding Papilionidae in the genus *Battus.* Other black, red, and white *Heraclides* females and males are thought to be Batesian satellites on the Mullerian mimicry rings around *Parides* (see *P. iphidamas,* #17), *Dysschema jansonis* (Arctiidae), and members of Pieridae, Nymphalidae, and Riodinidae, among others. But the males of *H. astyalus, H. androgeus, H. torquatus,* and both sexes of *H. cresphontes* and *H. thoas* bear brilliant yellow, large, flashy spots and stripes on a black background. These species have evolved in the direction of bright yellow, fast-flying Pieridae, presumably for the same reasons, signaling, "I am too fast and agile for you to catch me," or else they are models in a huge, unrecognized ring of largely inedible yellow-to-white Mullerian mimics. It is reasonable to ask why the males of some species belong to one mimicry complex and the females to another. The answer, which probably is correct, is that the males live a life of chasing females, sipping nectar at flowers, and sucking on ion-rich mud. They are thereby occupying one part of the prey-predator regime. On the other hand, females live a life aimed at locating sites for the deposition of eggs with some time spent sipping nectar from flowers. Thus, the females occupy a different, though partially overlapping, part of the prey-predator regime than the males, and in this context they have evolved to be participants in different mimicry complexes.

ADULT VOUCHER: 02-SRNP-15872; JCM
CATERPILLAR VOUCHER: 84-SRNP-1227; DHJ

76. *AGRIAS AMYDON* – NYMPHALIDAE

It is hard to think of wings that are any gaudier than the bright blue and red upper side and underside (shown in the gallery) of *Agrias amydon.* This immediately begs the question of why any butterfly would be so ostentatious? Or, put in another context, what selection process pushed for such attention-grabbing coloration? An easy answer is to note that all the species of the genus *Agrias* (only two species occur in Costa Rica, but there are many more in South America) and its near relatives (*Prepona* and *Archaeoprepona*) are large, diurnal, very fast flying and agile, generally cryptic on the underside (see *Archaeoprepona demophon,* #79 and *Prepona laertes* #80), and very alert to approaching animals, such as a large bird. They are very likely to attempt to flee from birds by flying fast and displaying a color pattern associated with fast flight. Since these butterflies have been around for millions of years, or so their genetic differences tell us, at some time in the long, unrecoverable past there was a particularly acute selection event for the ostentatious morph. It was influenced by some predator that was foiled by this appearance and this method of escape. This led to evolutionary radiation over the ages, resulting in many species of the butterfly exhibiting the basic pattern of the successful phenotype.

76

Such a commentary begs the question of whether this bright coloration is retained today primarily, or ordinarily, or only just enough through what is loosely termed phylogenetic inertia. In this case, a part of being brightly colored is simply that you have the genes for it and there is relatively little selection against them, with the trait of bright colors retained in the population through the failure of a dull morph to survive (predation?). Intertwined into such an explanation is the concept that the trait of bright colors is sufficiently complex and linked with other character states, features that are perhaps under more immediate selective maintenance. A mutation that renders a butterfly dull is removed from the gene pool for other reasons. What might these other attributes be? The most obvious is that it is likely, but unknown, that once in possession of a bright "don't bother with me" flag, that flag itself became an essential part of the butterfly's courtship, a sort of "peacock tail" possessed by both sexes and used by both sexes to evaluate the worthiness of the opposite sex. In this context, a dull individual would be selected against not only because it did not have the right clothes for the nuptial dance, but also because one gender would reject the other based on hue, tone, and brightness of color as a measure of whether the offspring will inherit other favorable features. Puzzling out the natural history of the bright colors of *Agrias* and other flashy, ostentatious, diurnal butterflies could easily become a tangled discussion. Science is not lacking for theoretical concepts. What is lacking is the empirical and experimental documentation of natural history, a most appropriate process for improving our understanding of how Lepidoptera conduct their daily lives.

ADULT VOUCHER: 04-SRNP-32315; JCM
CATERPILLAR VOUCHER: 04-SRNP-32315; DHJ

77. *URBANUS BELLI* – HESPERIIDAE

The roadside collector of butterflies can hardly escape encountering the long-tailed and various species of brown *Urbanus* in all the ACG ecosystems. They seem to be everywhere, visiting flowers, chasing each other, and poking among the herbaceous plants close to the ground in search of oviposition sites. They are yet another conspicuous butterfly morph that is ignored by the fly-catching birds perched on the fences, branches, and telephone lines. A number of *Urbanus* species showing this basic brown adult morph can be reared from a variety of caterpillar food plants, including Fabaceae: *U. simplicius, U. dorantes, U. proteus, U. evona, U. esmeraldus, U. esta,* and even *Chioides catillus* and *C. zilpa*; Poaceae: *U. teleus;* Urticaceae: *U. pronta;* and Asteraceae: *U. belli.* Definition of the latter species was thought to be clear. One day John Burns called from the Smithsonian Institution and said that he had discovered that one of the

77

U. "belli" specimens was *U. viterboana,* a species with slightly longer tails and somewhat more and denser blue-green on the upper side. Up to that moment, the few reared specimens of *U. viterboana* hailed from one small area halfway up the west side of Volcán Cacao, in the vicinity of the Estación Cacao rearing barn. This finding did not threaten the hegemony of *U. belli* caterpillars found in the ACG Asteraceae throughout the lowlands from deep rain forest to the driest far western end of the Santa Elena Peninsula.

Our confidence in the identity of *U. belli* was premature. As a matter of thoroughness, we decided to DNA barcode all of our reared specimens of *U. belli.* Its ACG hegemony went down the drain. Most inconveniently, there appear to be three *"U. belli":* one in deep rain forest (Rincón Rainforest and Estación Caribe), another occurring everywhere in the ACG, and the third mostly in rain forest (e.g., Sector Pitilla and Sector San Cristobal) but sporadically in dry forest as well. To date we cannot distinguish the three by anything but their DNA barcodes, as John Burns can find no distinctive differences in their genitalia. Their caterpillars look superficially the same and the food plants are roughly the same. The particular individual that we show has escaped barcoding to date, so we do not know whether it is a true and common *U. belli* (whatever that is, now that everything is questionable), or what we are currently calling the more rare species found primarily in the rain forest, *U.* BURNS02.

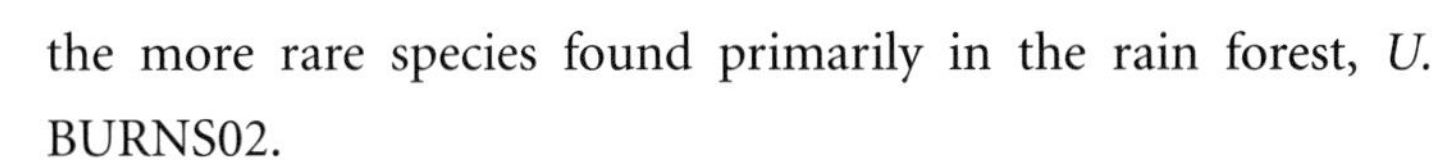

ADULT VOUCHER: 93-SRNP-3098; JCM
CATERPILLAR VOUCHER: 04-SRNP-15150; JCM

78. *ASTRAPTES TALUS* – HESPERIIDAE

Astraptes talus was described by Cramer in 1777 when he named it *Papilio talus,* which provides a historical note on how the higher level taxonomy of butterflies has developed over the past two centuries. Today *Papilio* is a small group of swallowtail butterflies (Papilionidae) that are evolutionarily distant from a skipper butterfly in the Hesperiidae. However, in the first years following the start of modern taxonomy (1758 to be exact), nearly all butterflies known at the time were placed in the genus *Papilio.* The other note of historical interest in the binomen of *A. talus* is the ancient age (230 years ago) of the specific epithet. Apparently this brown and iridescent green butterfly was common enough to have been among the first sets of butterflies brought back to the Old World by Neotropical explorers. Even today, turn someone loose with a butterfly net on a Central American roadside through lowland rain forest and *A. talus,* an obvious medium-sized skipper, is likely to be among the first captured.

Adult *A. talus* are part of a common pattern—brown with

iridescent blue-green highlights—that adorns so many species of medium-sized tropical pyrgine skippers, especially in the genus *Urbanus* (see *U. belli,* #77). However, *A. talus* stands out from the others, thereby allowing easy identification, by lacking tails on the hindwings. It has as many green scales on the underside of the body and wing bases and even the palpi as it has on the upper side. This is a very green butterfly.

The large and brilliant yellow-ringed black caterpillars feed on the mature foliage of several species of the huge, woody vine *Mucuna* (Fabaceae), which can sprawl over many hundreds of square meters of roadside vegetation next to the rain forest. *Mucuna* are famous for long, pendant inflorescences that are pollinated by bats and birds and for its (sometimes) highly urticating large pods, one of the sources of large, attractive, hard bean seeds found washed up on tropical beaches ("ojo de buey" as they are called). The caterpillars live solitarily in nests of *Mucuna* leaves, lightly silked one onto the other, and are probably Batesian mimics. At times, there can be hundreds of larvae on a single vine. After defoliating their food plant, the caterpillars wander off to pupate on neighboring vegetation, a behavior that can give the false impression that they feed on many other species of plants.

ADULT VOUCHER: 02-SRNP-28885; JCM
CATERPILLAR VOUCHER: 02-SRNP-28600; DHJ

79. *ARCHAEOPREPONA DEMOPHON* – NYMPHALIDAE

Incidentally, more than a century ago some taxonomist played a bad joke regarding etymology. There is *Archaeoprepona demophoon,* which eats *Ocotea veraguensis* in the dry forest but also ranges into the rain forest where its caterpillars feed on many species of Lauraceae, and *A. demophon,* which we show here. The confusion created by these two species epithets has resulted in an unholy mixup in food-plant records, most of which are useless because people said one of the species when they meant the other, and because well-meaning editors and readers assumed that "*demophoon*" and "*demophon*" were simply alternative spellings of the same name.

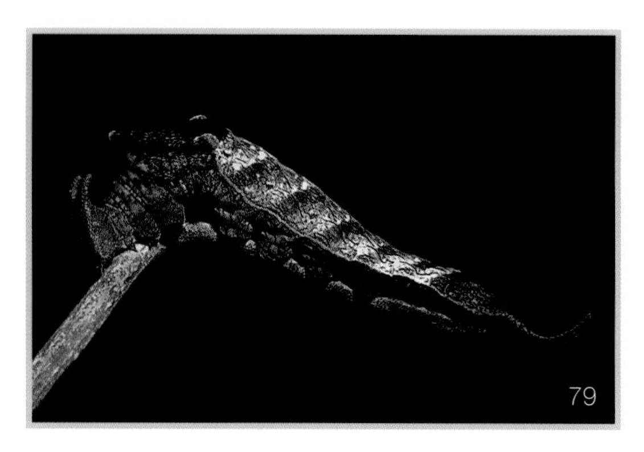
79

The silvery gray underside of *A. demophon* blends marvelously with the bark of a tree, resembling a torn flap of bark. If feeding, its bright pink tongue continues to probe the crack in front of it, seeking to suck up the fermenting sap and probably the yeasts inside. While, on the one hand, it often appears to be drunk on the alcohols (and other products?) that we all know yeasts are so good at making as part of their defensive arsenal against other microbes and even big animals (Janzen 1977), on the other hand, it may be that *A. demophon* has nerves of steel as it watches your hand approach. It may even be that these drinkers of fermenting sap and yeast would be very reasonable organisms to use in the search for alcohol-degrading enzymes, which would help us understand those aspects of physiology adapted to the ingestion of alcohol molecules.

The butterfly will remain quite still, watching with its big compound eyes and not flinching, until it decides that you are just too close. When you finally touch the wings, or just before, *A. demophon* will launch into an extremely fast, almost violent flapping, sailing flight that displays most prominently the two intense, iridescent blue to turquoise bands on the upper side of its wings. This pair of flags, as in iridescent blue *Morpho amathonte* (#34) and *Doxocopa,* says, "I know you see me and I see you, but don't even bother to try to chase me." This is also the message the white-flagged rear of white-tailed deer and cotton-tailed rabbits send as they disappear into the woods.

The food plants of *A. demophon* caterpillars stand out as quite different and more diversified than those of the other ACG caterpillars in this complex: *Agrias* feeding on *Erythroxylon,* the two species of *Prepona* each on Chrysobalanaceae and Fabaceae, and *A. camilla* and *A. demophoon* on Lauraceae. We have found *A. demophon* caterpillars feeding on at least fifteen species in nine plant families. It is particularly striking that none of these are species of plants fed on by the other five species mentioned above.

ADULT VOUCHER: 98-SRNP-6820; JCM
CATERPILLAR VOUCHER: 02-SRNP-2966; DHJ

80. *PREPONA LAERTES* – NYMPHALIDAE

Anyone who has put out a butterfly trap baited with fermenting bananas in the ACG forest knows *Prepona laertes.* This butterfly has striking false eyespots on the underside along with two gaudy, iridescent multi-blue stripes and a remnant of an eyespot on the brown upper side. There is no hint of mystery in the silvery-to-beige underside. Its complex pattern looks like crevice margins in the tree bark it commonly sits on while probing and sucking with its pink tongue in a yeast- and ferment-filled wound. This species is the smallest of the five known species in the ACG *Prepona-Archaeoprepona* cluster of charaxine nymphalids and is present in any lowland habitat that still has trees and an understory. It ranges from tropical Mexico to Venezuela and Peru, and is known by a variety of subspecific names that have been applied to it over this range, revealing the inclination of taxonomists to baptize each morph of a spectacular butterfly with its own scientific trinomen.

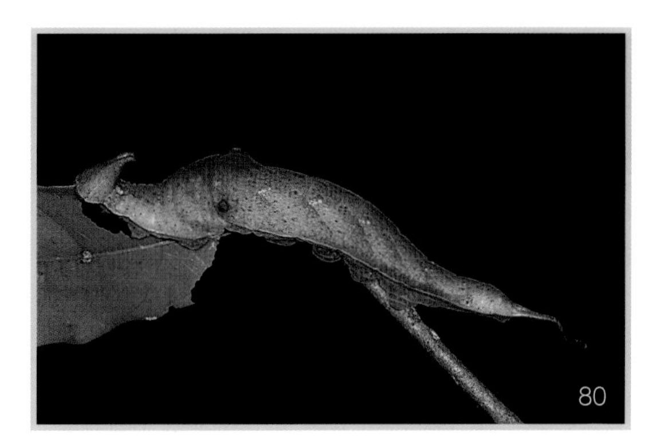
80

Taxonomic revisions are an essential part of ecological and evolutionary studies because all is not as it seems and new information will either strengthen or change existing classifications. Our first clue that the taxonomic stature of *P. laertes* had change written into its future was finding the leaf-mimicking caterpillars feeding on *Hirtella racemosa* and *Licania arborea* in the Chrysobalanaceae, and on six species of Fabaceae in the genera

Inga, Zygia, and *Andira.* There are no other species of ACG caterpillars that feed on just these two sets of food plants. These two plant families are not at all closely related, and they are not known to be similar in their defensive chemistry. The second clue came from a perceptive South American lepidopterist who wrote that he thought that some detail of color of the upper side of the *P. laertes* specimens shown on the Web site suggested that there might be two species. The third clue came to us when we DNA-barcoded the ACG specimens and found that what we had been calling *P. laertes* in the ACG is definitely two biological entities, one group feeding on Chrysobalanaceae and the other feeding on Fabaceae. The male adult we show is *P. laertes* as it is defined by the scientific community at the moment of this writing and is the present name of record. This male was reared from a caterpillar collected in 1993 while feeding on *Inga vera.* We hesitate to put a temporary interim name to it because we do not know which barcode corresponds to the group that matches the type specimen of *P. laertes.*

ADULT VOUCHER: 93-SRNP-2759; JCM
CATERPILLAR VOUCHER: 04-SRNP-14567; JCM

81. *APHRISSA STATIRA* – PIERIDAE

"The butterflies are merciless today," said a cartoon burned into memory, and so they were one late June. Looking west down the channel of the paved road of the official ACG entrance at the Casetilla Entrada, there was an aerial river of large, bright yellow butterflies pouring out of Sector Santa Rosa's dry forest at the end of their first generation of the year. They passed in ones to twenties every few seconds, flying one to four meters above the ground, males and females, the children of parents that had arrived at the ACG dry forest a month earlier to lay their eggs on the new foliage marking the beginning of the rainy season. One of the more common members of this mass of large Pieridae was *Aphrissa statira,* along with five species of *Phoebis* and several other genera.

From the ACG dry-forest viewpoint, *A. statira* and many others are largely univoltine. That is to say, they have one generation each year, conducted at the time when the food is at its highest quality and the parasite and predator regime is at its least intense (Janzen 1987a, b; 1988a, b). But instead of being dormant as a pupa or a sexually repressed adult, they leave. They then have another generation or generations away from the ACG, elsewhere in the wetter parts of Costa Rica. The great-grandchildren, or yet further descended, return a year later to repeat the cycle. Centuries ago when the ACG was one unbroken stretch of old-growth forest from the Pacific coast to the Caribbean lowlands (as it will come to be again some day), the migration of large yellow pierids with the first rains, their single dry-forest generation, and their massive exit to wetter areas to the east, was probably very cleanly performed. Today, a few individuals of some species stay behind, though not *A. statira,* to have one to three minimalist rainy-season generations on the small ocean of food plants, especially *Senna* and *Cassia* (Fabaceae), in the early successional edges of dry-forest roadsides and pastures. However, *A. statira* is a specialist on the very new foliage of *Callichlamys latifolia,* a long-lived, huge vine from old-growth forest, and the secondary successional woody vine *Xylophragma seemannianum,* both Bignoniaceae. As a food specialist and a selective feeder based on foliage age, *A. statira* restricts itself to a single generation before leaving for points unknown. The foliage-green caterpillars are abundant and evident to any foraging bird because they perch on both sides of a leaf and because their damage is often the first to be sustained by the new and large leaves of their food plants. However, population levels can be enormous, so it seems there may be safety in numbers. The forest at this time of the year is rich in green edible caterpillars. Such an abundance of prey may divert the attention of potential predators to a mix of species and perhaps also serves to satiate the predators.

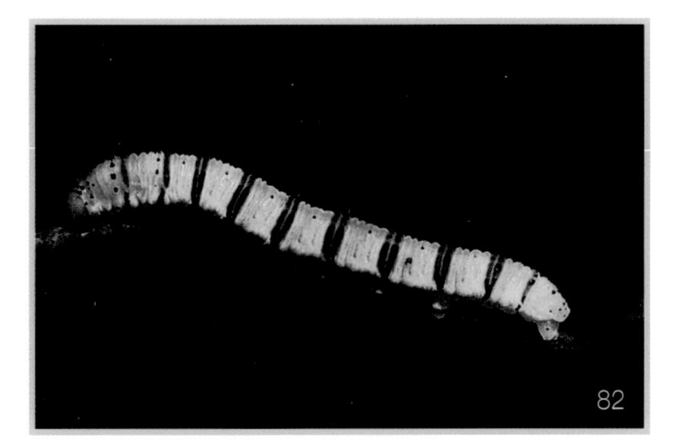

ADULT VOUCHER: 84-SRNP-232.03; JCM
CATERPILLAR VOUCHER: 03-SRNP-12737; DHJ

82. *PHOEBIS PHILEA* – PIERIDAE

Pieridae is the family of the sulphur butterflies and the whites that are so prominent on any landscape, tropical or extra-

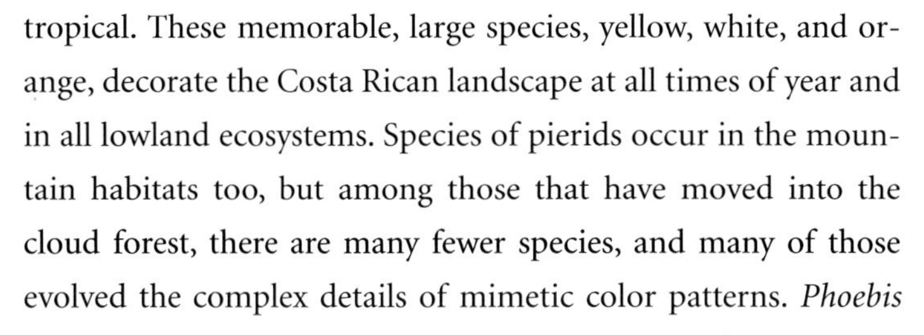

tropical. These memorable, large species, yellow, white, and orange, decorate the Costa Rican landscape at all times of year and in all lowland ecosystems. Species of pierids occur in the mountain habitats too, but among those that have moved into the cloud forest, there are many fewer species, and many of those evolved the complex details of mimetic color patterns. *Phoebis philea* is one of three pierids we portray, the others being *Aphrissa statira* (#81), a classic large, yellow pierid (as is *P. philea*), and *Dismorphia amphiona* (#72), a black and orange, tiger-striped heliconiine look-alike.

To perform an experiment that tells you something about the biology of *P. philea* and many other bright white and yellow pierids, collect a liter or so of urine and pour it onto the riverbank mud or wet roadside in a relatively secluded place, yet fully insolated. The seclusion is to ensure that passing vehicles and people won't disturb the butterflies that will respond to the bait. Give the site a day to percolate, come back during a hot, sunny midday, and you may well find several species of pierids, including *P. philea* males (with a bright pink-red overlay on their yellow hindwings and a rosy wide bar on the front wings), perched on the ground with their tongues stuck firmly into the mud or sand. At this point, stealthily get down on your hands and knees and crawl up to them slowly. Avoid making quick movements, because the butterflies are easily startled. Keep as low a profile as possible, because they seem to notice movement against the sky better than movement against the ground and vegetation. When you are close enough, with your camera held out in front, watch both the front and the rear of the butterfly carefully. The tongue will probe and move and probe and move and eventually find a place to stay. Soon thereafter a drop of water will fall from the tip of the abdomen, and another, and another. It will be only a male doing this. He is passing as much liquid as he can through his gut while filtering out the sodium, and perhaps calcium, ions. He concentrates these ions and includes them as a "nuptial gift" in the sperm packets that he delivers to the female. It may well be that the amount of sodium in a given sperm packet determines whether she digests it or allows it to fertilize her eggs. In this sense, she uses the sodium as part of the behavioral-physiological process of producing eggs. Since the female digests most of the sperm packets that she receives through her multiple matings and uses the amino acids contained therein in egg development, the sodium content in the sperm packet, in part, determines if a given male merely passes protein into the next generation or his genes.

ADULT VOUCHER: 82-SRNP-4.1; JCM
CATERPILLAR VOUCHER: 82-SRNP-4; DHJ

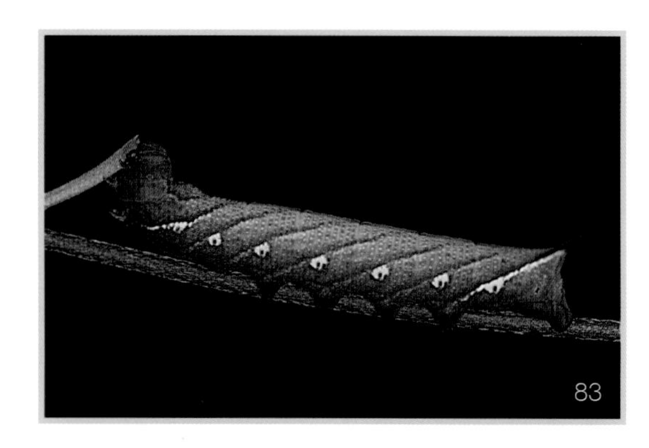

83. *AELLOPOS CECULUS* – SPHINGIDAE

When a sphinx moth shifts from nocturnal habits into carrying out its adult activities in full daylight, it is either a physically small species whose near cousins are small as well, or it has evolved into its diminutive stature from a lineage of larger species, as attested to by the small size of the few fully diurnal sphingids scattered across the globe. In other words, small species are not necessarily diurnal, but diurnal species are necessarily small. *Aellopos ceculus* is such a small, diurnal sphingid species found in the ACG lowland rain forest. A large individual has a wingspread of four to five centimeters, which puts it in the bottom quarter of the size distribution of ACG sphingids. All three of the *Aellopos* (*A. titan, A. fadus,* and *A. clavipes*) in the ACG dry forest follow the rule and are also fully diurnal. They may occasionally occur within or pass through the rain forest. However, it is *A. ceculus* that the naturalist encounters buzzing frantically, like a lightning-fast, yellow-banded, minute hummingbird, among the rain-forest roadside vegetation looking for oviposition sites on *Sabicea villosa* and *Sabicea panamensis,* its caterpillar food plants. These are somewhat herbaceous vines in the Rubiaceae. Despite being a moth of the bright sun and relatively dry weather found in early successional, disturbed sites in the rain forest, *A. ceculus* does not penetrate the adjacent dry forest, even during the six-month rainy season. This absence is quite puzzling, since there are many tens of species of dry-forest Rubiaceae that could serve as potential caterpillar food plants. But, and this is the best clue, *Sabicea,* like the *A. ceculus* caterpillars that eat it, does not enter the ACG dry forest either, presumably because of unfavorable environmental conditions. Furthermore, *A. ceculus* does not extend up into the ACG mountains and cloud forests. Again, the reason *A. ceculus* prefers the rain forest may consist of nothing more than the absence of its caterpillars' food plants.

The bright yellow hindwing patches and their orange continuation onto the sides of the abdomen are ever so visible as this chubby moth buzzes from plant to plant in broad daylight. Presumably the bright colors play the same role as the bright white abdominal ring on the other three *Aellopos* in dry forest. It is a reasonable inference that at some time in the history of *Aellopos,* there were birds that focused on the bright spots of fast-moving potential food and then lost the food when the spot disappeared, either by landing and covering the spot with the wings, or dodging behind something. This permanently visible diurnal marker may be as important in courtship as in escape from birds.

All four species of *Aellopos* have an extremely fast wingbeat that results in a buzzing flight with an audible resemblance to something between a large carpenter bee and a small hummingbird. Curiously, these moths also have a hard and extremely slippery body. Even if a foraging bird were agile enough to get within striking distance and make contact with an *Aellopos,* it would have a difficult time holding onto its prey, as would a monkey equipped with soft fingertips designed for picking up objects the size of an *Aellopos.*

ADULT VOUCHER: 03-SRNP-10394; JCM
CATERPILLAR VOUCHER: 04-SRNP-42562; JCM

84. *ENYO OCYPETE* – SPHINGIDAE

Enyo ocypete, though only occasionally captured at ACG light traps, has arguably the most reliably encountered and most easily located caterpillar of any sphingid during the rainy season in the dry forest. The same is true in the lowland rain forest. Its usual food plants are Dilleniaceae: *Curatella americana, Doliocarpus dentata, Davilla kunthii,* and *Tetracera volubilis* in dry forest; and *Tetracera hydrophilla* and *Davilla nitida* in rain forest. Occasionally caterpillars are found feeding on various species of *Cissus* (Vitaceae).

The adults, or at least the female adults, can be tracked seasonally within the ACG by the presence of their early-instar caterpillars. There can be young *E. ocypete* caterpillars on the new foliage of the small tree *Curatella americana,* growing in arid land, as much as one month before the rains, which begin in mid May in the ACG dry forest. No adults come to a light trap at this time, so it may be that courtship occurs at another time. The females may be using sperm obtained in matings months earlier or in the wetter parts of the ACG. Then as the rains arrive, so do *E. ocypete* adults from either local moist areas such as riparian vegetation or from the rain forest in the eastern part of the ACG. Females oviposit throughout the dry forest and, judg-

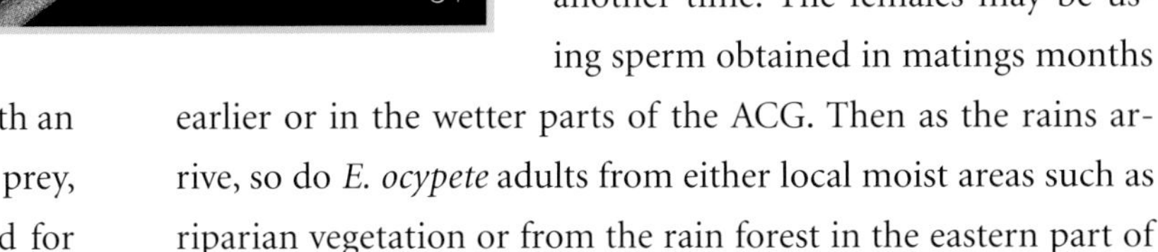
84

ing from the ages of the caterpillars during the first half of the dry-forest rainy season, they tend to oviposit over a period of about six weeks on the new foliage. After a single generation, and perhaps a second, the newly eclosed adults appear to leave, even though the rainy season is only half over, and abundant caterpillar food is still available. Apparently the adults fly to the rain-forest side of the ACG, an assumption based on the observation that in the rain forest there are caterpillars present at a much lower density throughout the year. It is impossible to know if there is a continuously resident population in the rain forest or if another migrant population moves back and forth between the wet and dry forest according to the seasons. We suspect the former, but the latter could be happening. There is no hint in the genetic signals obtained by DNA barcoding that there are two distinct populations.

The males and females of *E. ocypete,* as is the case with the other species of *Enyo* as well, are strikingly different in wing coloration, and even in the details of the shape of the forewing margin. The members of *Enyo* are the only species of ACG sphingid in which the sexes can be determined from a distance, often while they are in flight. Given the powerful and complex flight ability of sphingids, having different shapes and colors of the forewings suggests that the sexes may have different flight behaviors and perhaps may even visit different species of flowers.

ADULT VOUCHER: 03-SRNP-15719; JCM
CATERPILLAR VOUCHER: 05-SRNP-47433; JCM

85. *AUTOMERIS METZLI* – SATURNIIDAE

85

Automeris metzli entered the ACG dry-forest caterpillar inventory project as a road kill. In 1981, a thoroughly squashed, huge, urticating, spiny caterpillar was found on the asphalt road in the Bosque Humedo. No adult had ever come to the light traps that seemed appropriate to represent it. Four years later, a huge female of *A. metzli* was caught in a light trap in the same forest, and we wondered if the two went together. However, her offspring, acquired from the eggs she had laid in a plastic bag, refused to eat the typical *Automeris* food plants, *Inga* (Fabaceae) and *Luehea* (Tilliaceae), and died of starvation. The mystery remained unsolved for four more years. In 1988, a second female was captured, but her eggs were sterile. Then, in May of 1989, the mystery of the eight-year-old road kill was finally solved. The first parataxonomist course (Janzen 2004) was billeted in the oak forest at Finca Jenny (Sector Santa Rosa). Two weeks after the rains began, we noted that there were masses of a huge, gregarious, urticating, hemileucine saturniid defoliating the oaks overhead, raining buckets of frass pellets from the canopy. At last the caterpillar of *A. metzli* was found. A second generation followed in late August.

Not only is *A. metzli* the largest of the five species of *Au-*

tomeris caterpillars in the ACG dry forest, but it is also the only species that remains gregarious throughout all instars and even spins its cocoons side by side under loose bark or litter at the base of the tree. As an oak feeder, *A. metzli* gives the impression of being a dry-forest species. It is common in the lowland oak stands in the ACG dry forest, and it feeds on several other species of dry-forest trees, namely, *Cedrela, Hymenaea, Ochroma, Dalbergia, Trema,* and *Tabebuia*. But it also occurs in ACG rain forest, where it lives with a similar-looking species, *A. exigua*, which is restricted to the rain forest. The adults of *A. exigua* have slightly more blunted tips to the forewings, slightly smaller false eyespots on the hindwings, and some very subtle differences in the beige-brown patterning of the forewings (see adults at http://janzen.sas.upenn.edu). The caterpillars of *A. exigua* are also gregarious throughout their lives, and the hundreds of caterpillars resulting from one oviposition event can defoliate an entire large rain-forest tree (just as a group of sibs of *A. metzli* can defoliate a medium-sized *Q. oleoides* tree). This pair of species even shares species of parasitoids, and were it not for the painstaking identification of ACG Saturniidae by Claude Lemaire (2002), this is a pair of species that we would likely have lumped together under one name until DNA barcoding proved us wrong.

ADULT VOUCHER: 99-SRNP-10818; JCM
CATERPILLAR VOUCHER: 01-SRNP-2315; DHJ

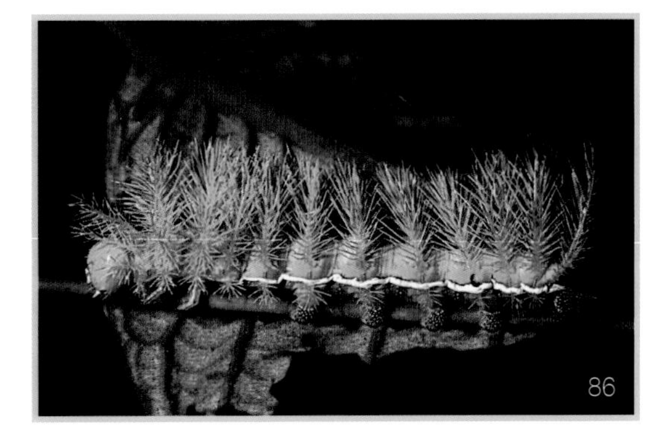

86. *AUTOMERIS PHRYNON* – SATURNIIDAE

When you meet your first *Automeris phrynon* while it is sitting on a light-trap sheet (this moth has yet to be met in the wild in the ACG), your reaction may be, "Oh, yet another species of *Automeris*." But when she spreads her wings to display the false eyespots on the hindwings, as all *Automeris* do when molested, you realize that there is something unusual about this species. The eyespots are small, weak, and not well developed. Then, after a while, you may notice that you have never seen a male at the lights, yet males in other species of *Automeris* are common at lights. It took the expertise of Bob Marquis to figure it out while he was at the OTS La Selva Biological Station in the lowland rain forest, 100 kilometers to the east of the ACG (Marquis 1984). Bob reared a clutch of *A. phrynon* green urticating caterpillars, resembling those of *Automeris tridens* (Miller et al. 2006), and startled all of us by discovering that the male does not look at all like a classic *Automeris*. Its false eyespots are reduced even more than in the female, and about the only striking decoration on its dead-leaf wings is a large, bright yellow area on the underside of the forewings. But that is not the real discovery. He found the males flying vigorously in broad daylight, responding to females emitting their pheromones between 9 and 11 AM. The male wingbeat is so fast that it renders the moth just a blur with a bright yellow hue in the center. Is it a mimic of pierid butterflies, simply a yellow equivalent to the bright blue "don't bother to try" flashes of *Morpho* and *Archaeoprepona,* or equivalent to the yellow hindwings of sphingids and *Gonodonta* noctuids? That remains to be worked out, but it is certainly an eye-catching aspect of this strange, diurnal saturniid.

What selective force moved the *A. phrynon* chemical courtship into the full daylight hours? Diurnal courtship is a behavior that is expected in cold northern climates or high elevations, as with *Arias inbio,* a Costa Rican high-elevation saturniid that looks like a pierid butterfly (Lemaire 2002). It is not expected in the hot lowland tropics. The only relevant explanation, given our dismal knowledge of the natural history of these beasts, is that it occurs in a rain-forest world containing five to eight species of *Automeris,* all with overlapping generations. On any given night of any season, it is possible that all of them are chemically courting. It is likely that congeneric pheromones have much in common, and one way to find a portion of chemical-free airspace that is not contaminated with males and chemicals of the other species is to shift the circadian rhythm for mating into the daylight hours. It is likely that this is also why *Rothschildia erycina* (#94) is courting in the hours just before and at dawn, while its congeners, *R. lebeau* and *R. triloba,* court during the long, dark night.

ADULT VOUCHER: 05-SRNP-1067; JCM
CATERPILLAR VOUCHER: 02-SRNP-3156; DHJ

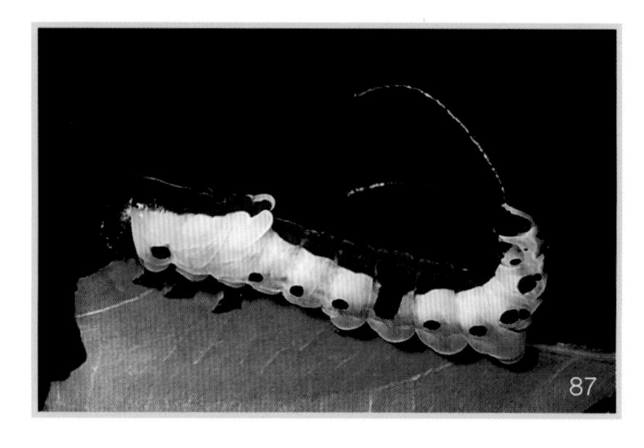
87

87. *CERURA DANDON* – NOTODONTIDAE

The pattern of the adult of *Cerura dandon,* stark black and white, but blurry, is easily interpreted as being cryptic on a lichen-covered tree trunk, but maybe that interpretation is too simple. There is another thesis. Skunks are black and white too. In fact, black and white is a common aposematic coloration in the world of colorblind predators or predators that operate in the dusk and night when colors are turned to black and white by lack of direct sunlight. The possibility of aposematism (only to be truly determined by experimentation) arises from the behavior of the adults when handled and from the caterpillar colors.

When the adults are handled, they tend to become inactive and curl the abdomen, accentuating the black and white rings and the overall black and white pattern of the body. Their inactivity may well protect against the "kill it if it moves" reflex that comes naturally to a predator for whom fast movement signals possible escape of the prey. Such inactivity stands in strong contrast to the black and white bark-patterned, and presumably quite edible, Geometridae and Noctuidae in the ACG. These moths frantically fly away when threatened.

The brilliant red and yellow of the last instar do not conform to the usually cryptic colors and patterns of so many of the ACG notodontid caterpillars. Whereas notodontids are generally viewed as highly edible, escaping through camouflage or mimicry of vertebrate eyes, there are a few ostentatious species.

Certainly, *C. dandon* is one of them. However, before accepting a hypothetical aposematic function for the caterpillar colors, consider the fact that the food plants, Flacourtiaceae of old, presently placed in the willow family (Salicaceae), are not noted for possessing toxins that could easily be sequestered by the caterpillar or modified into nasty defenses. Still, an acutely nasty chemical does not necessarily have to be present. Perhaps something like the active ingredients in aspirin, salicylic acid (a natural ingredient of willow tissue), might play a role in the defensive biology of *C. dandon* adults and caterpillars.

ADULT VOUCHER: 95-SRNP-9847; JCM
CATERPILLAR VOUCHER: 05-SRNP-47385; JCM

88. *CERURA RARATA* – NOTODONTIDAE

Just as there are many ways to visualize a dead leaf instead of a moth's front wings, there are also many ways to visualize a patch of dirty lichen. *Cerura rarata,* to be contrasted with *Cerura dandon* (#87) and *Hypercompe icasia* (#53) on this point, represents one of those ways. But there is a fine line between fact and conjecture in this case. White-with-dark is a much less common color pattern for day-sleeping moths, an appearance and corresponding behavior that is scattered across a variety of moth families. This pattern is often found in moths that are suspected of being aposematic rather than cryptic, but with *Cerura* we simply do not know. The forest is full of white to pale bits of this or that, and white to pale reflections as well, so white, especially when broken with dark lines and patches, can be quite cryptic. However, when this white-with-dark pattern is layered onto a pair of wings that can be symmetrically the same if viewed from above, there is the risk of inviting closer inspection by a potential predator, perhaps more risk than the humdrum brown leaf morph undertakes. When disturbed, both species of *Cerura* feign death while curling and exposing their black and white-ringed abdomens, yet another signal to the potential consumer that if eaten, these moths may induce an episode of vomiting. Incidentally, if we are to search for the selection for these color patterns, we need to search far back in time. There are species of *Cerura* in the Old World tropics, long separated from the Neotropics, with essentially identical wing color patterns.

Cerura rarata generated a surprise for the ACG inventory project. When DNA barcoded, the sequences from many specimens fell into two distinctive clusters, suggesting that what appears to be so nicely a single ACG species may in fact be two species differing by as much as 6 percent. Although the genitalia of individuals in these two clusters have yet to be compared, the facies of the specimens appear to be identical. This is no surprise, because cryptic species are common among ACG Notodontidae and close inspection does find morphological, behavioral, caterpillar, food plant, and microgeographic differences between the siblings. However, a problem emerged upon detailed examination of the specimen records. There are what appear to be siblings in both of the sequence groups. Noting that the DNA barcode is inherited from the mother (matrilineal), something is awry if siblings have strongly different barcodes.

We can propose at least three hypotheses addressing the sibling barcode results. First, as implied above, the species may be highly polymorphic in such a manner that a single mother may have offspring bearing either morph, but this seems most unlikely since the DNA barcode sequences are mitochondrial and therefore inherited only from the mother. Second, the species may be gregarious in its oviposition habits, such that when a cluster of caterpillars is found on a single plant, they are from multiple females. Other Lepidoptera are known to seek oviposition sites used by conspecific females, though such "safety in numbers" strategies are somewhat unusual. Third, and the most likely explanation, is that certain individual specimens of *C. rarata* were contaminated with the DNA of another species, and in the sequencing process, the barcode of the contaminant was captured rather than the barcode of the target specimen.

ADULT VOUCHER: 98-SRNP-9917; JCM
CATERPILLAR VOUCHER: 02-SRNP-24403; DHJ

89

89. *AZETA RHODOGASTER* – NOCTUIDAE

If you see a noctuid moth in the ACG with a red abdomen, it is *Azeta rhodogaster.* As its species epithet suggests, *rhodo-* is "red" and *-gaster* means "abdomen." There is no other moth like it. The bright red abdomen also lets us know that this species is more diurnal and more inclined to fly when a predator approaches than are the many hundreds of other species of ACG Noctuidae. It perches exposed on the foliage, watches for movement with its large compound eyes, and then flies away when approached. It is very likely that the red abdomen functions in the same manner as the yellow hindwings on so many noctuids. A searching bird zeroes in on the red as the moth is in flight, but the red abruptly disappears when the moth alights, often under a leaf. There is no evidence that the red abdomen serves as an aposematic flag that warns of a toxic meal on wings.

The caterpillar of *A. rhodogaster* also steers the researcher away from assigning aposematism and toxic chemicals as a likely defense mechanism. The pale, greenish-yellow and black dotted caterpillars are food-plant specific to *Gliricidia sepium* (Fabaceae), a small to very large native tree in the ACG dry forest. This dry-forest tree has been introduced around the global tropics as a living fencepost and in the ACG is present in both dry and rain forest. When this tree was planted as the support for barbed-wire fences on the rain-forest side of Costa Rica, *A. rhodogaster* was just one of the many dry-forest species (also *Epargyreus* BURNS02 in the Hesperiidae and *Dasylophia basitincta* in the Notodontidae) that moved from dry- to rain-forest pastures, following its food plant.

Caterpillars of *A. rhodogaster* can occur in very high numbers and can completely defoliate their food plants if the plant is a sapling or a young treelet. At these times, it is easy to collect many of the caterpillars to see what parasitoids are using them. Apparently it is also easy for at least one parasitoid species to find them. A medium-sized, somewhat generalist tachinid fly, *Eucelatoria armigera,* killed at least 40 percent of the caterpillars in one sample of 446, mostly last instars collected during a population explosion from just a few *G. sepium* treelets in ACG dry forest. Interestingly, no parasitoids are specialists on *A. rhodogaster* caterpillars.

ADULT VOUCHER: 05-SRNP-18946; JCM
CATERPILLAR VOUCHER: 83-SRNP-140; DHJ

90. *PSEUDODIRPHIA MENANDER* – SATURNIIDAE

Not all saturniids that look gaudy in a museum drawer are like that when found in nature. In the field they may be cryptic, hanging by one leg and resembling a dead leaf instead of spreading their wings. However, take note that those that look like a dead leaf in a museum drawer do indeed look the same when seen in the field. As presented in our adult image, *Pseudodirphia menander,* with its brilliant reddish pink and orange, would appear to be some curious combination of these two observations. Despite their appearance, these are not aposematic colors when viewed at a distance. The rain-forest and cloud-forest world this moth occupies has many examples of red and pink leaves, especially among the new leaf flushes that are protected from ultraviolet rays in sunlight by red pigments, a sort of natural sunscreen. When *P. menander* is perched sloppily across the twigs and leaves in the canopy, it looks like a new pink leaf.

However, the defensive behavior of *P. menander* does not conform to that of a cryptic species. If prodded or grabbed, the moth does not flee. Instead, it pops its wings up and into a semi-spread position, inflates and curls its abdomen into a bright orange and black-ringed aspect, and goes into a catatonic state while releasing a strong, musky odor. This behavior and chemistry signal, "Leave me alone, or you will regret taking a bite." It is cryptic when undisturbed, but turns unpleasant and potentially dangerous when found. Having never watched a bird or monkey discover an adult *P. menander,* we cannot vouch for the effectiveness of this defense, but it is obvious it holds survival value. What we do know is that if the moth, in its defensive posture, is dropped into a swarm of *Eciton* army ants (vicious predators of insects of all sizes), there is an immediate ant-free area in the swarm with an untouched and unharmed moth lying in the middle. This effect on ants also occurs with *Dirphia avia* and *Periphoba arcaei* (see Miller et al. 2006), two other cryptic ACG hemileucine saturniids that, when threatened, remain in place while curling their black-ringed abdomens and emit a foul, repulsive chemical.

Were it not for the apparent cryptic nature of the moth, the caterpillar food plant, the food-plant chemistry, and adult coloration would certainly support the speculation that this is a species possessing aposematic traits. So, for the moment, eschew the dichotomy of cryptic versus aposematic and instead combine the two with some additional considerations. From afar the vertebrate eye does not detect the moth of *P. menander*. However, upon a chance close encounter, the red coloration warns the intruder to back off. If that signal is not perceived, then the chemistry of repulsion is employed to instill the relationship between color and a foul taste. The selection for the ability to repel predators is further reinforced by its effectiveness against invertebrates—ants in particular—which likely do not perceive the colors in either a cryptic or aposematic context.

The caterpillar of *P. menander* is, as with other ACG Hemileucinae, somewhat cryptic, somewhat aposematic, and above all extremely urticating. Its distinctiveness is that instead of being polyphagous, as are most hemileucines, it is a specialist on the foliage of epiphytic Araceae, plants that are well known for harboring nasty secondary compounds that strongly impact the vertebrate adventurous enough to bite into them. It may well be that in addition to self-made chemical defenses (the other two foul-smelling Hemileucinae mentioned above are experts at making their own defensive chemicals), the caterpillar of *P. menander* is also sequestering defensive chemicals from the aroid food plant and perhaps passing them on to the adult.

ADULT VOUCHER: 97-SRNP-1764; JCM
CATERPILLAR VOUCHER: 04-SRNP-60423; JCM

91. *PHOCIDES NIGRESCENS* – HESPERIIDAE

Flip over a specimen from the dorsum to the venter and back again, and you see the same thing. Once you have seen one side of *Phocides nigrescens*, you have seen both sides. At least eight ACG hesperiid species in four genera and two subfamilies show this type of wing color and pattern: *Jemadia pseudognetus*, *Jemadia* BURNS01, *Phocides belus*, *Phocides pigmalion*, *Phocides* WARREN01, *Parelbella macleannani* (#92), and *Elbella* BURNS01. Furthermore, the list could include *Myscelus assaricus michaeli*, which is *P. nigrescens* on the upper side and *Yanguna cosyra* (see Miller et al. 2006) on the underside. Perhaps the most fancy skipper of them all is *Tarsoctenus corytus gaudialis*, which occurs in the ACG, but whose caterpillar has not yet been found. The male looks like *P. nigrescens* and the female looks like the black-with-orange *Y. cosyra*. But the ACG is not the epicenter of the blue-white-black striped mimicry complex. The color plates in *Monograph of the Neotropical Butterflies* (Seitz), first published in the early 1900s and now outdated, shows that more than fifty species of hesperiids in many genera have converged into this defensive appearance. As with all the other bright flashing blue ACG butterflies—*Morpho*, *Archaeoprepona*, *Myscelia*, *Doxocopa*, *Memphis*, and others—

these colors are most likely saying, "Don't bother to try" to the aerial foraging predator. It is no dumb guess to suggest that these colors are tied into the courtship signals as well.

In contrast to the chemically defended, aposematic ACG butterflies, many species of which are encountered nearly everywhere, the *P. nigrescens* color pattern is not generally visible to the casual observer. Many of these butterflies have never been seen as wild, free-flying adults nor collected with a net in the ACG, despite the rearing of several thousand of their caterpillars. The few that have been seen, along with other medium-sized Hesperiidae, were sucking nectar from *Inga* and *Pithecellobium* flowers in tree crowns. This is a hint that the adults live in another world, high in the canopy, even though their caterpillars live down among us.

Caterpillars of *P. nigrescens* feed on eighteen species in four genera of ACG Myrtaceae and are part of a different mimicry complex than the adults. The last instar is white, looking something like a bird dropping or a mold-covered, dead pupa, sporting a dark brownish red head with glaring, false yellow eyes. Dozens of species of hesperiid caterpillars share this appearance. The first four instars are dark reddish brown, semi-ringed with bright yellow, thereby falling into the large complex including many more species of ACG caterpillars that are essentially dark with yellow, orange, or red rings. Which species are the models, which are the mimics, and which benefit most from the birds' genetic hardwiring to avoid ringed objects (Smith 1975)? These questions are still unanswered. Nonetheless, the existence of this mimic-model complex cannot be denied.

ADULT VOUCHER: 04-SRNP-2822; JCM
CATERPILLAR VOUCHER: 99-SRNP-15140; DHJ

92. *PARELBELLA MACLEANNANI* – HESPERIIDAE

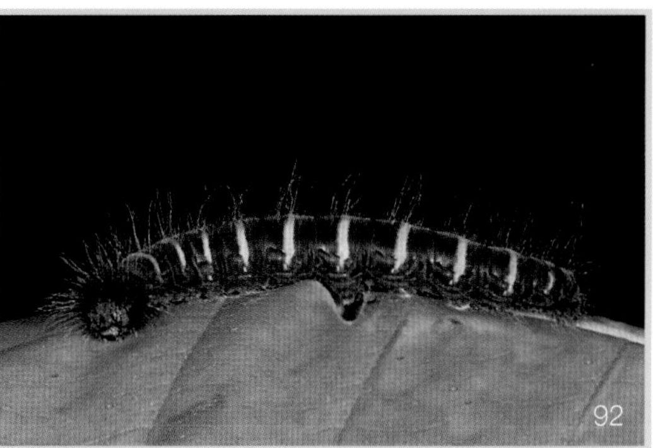

There is little chance of failing to notice the existence of mimicry complexes in the Neotropics. Mimics are everywhere. For instance, compare *Parelbella macleannani* with *Phocides nigrescens* (#91), or for that matter the images of the other *Phocides, Jemadia, Elbella,* and *Myscelus* (go to http://janzen.sas.upenn.edu). Now imagine trying to compare them when they fly by at fifty kilometers per hour. This is what a bird likely sees—a medium-sized blur of black, white, and iridescent blue. The upper sides of these two species seem to be nearly identical, except for the details of the blue stripes and the form of the translucent white windows in the front wings, but the underside of *P. macleannani* is more conspicuously turquoise while the underside of *P. nigrescens* is the blue of the upper side. When in flight, they may be merely blue to a predator, but in the eyes of the opposite sex, they are probably as different as night and day.

These two butterflies are in different subfamilies of the Hesperiidae. *Parelbella macleannani* is in the Pyrrhopyginae, a subfamily of medium-large butterflies, not particularly species-rich with only twenty-five or so species in the ACG. The caterpillars are very hairy, including massive facial hairs, and very Neotropical (Burns and Janzen 2001). In contrast, *P. nigrescens* is in the Pyrginae, a subfamily with species represented by all sizes of butterflies, very species rich with at least three-hundred species in the ACG. The caterpillars are often naked and almost always with naked faces, and they can be found all over the world. Hesperiid butterflies have been with us for many millions of years. A guess is that the Pyrrhopyginae-Pyrginae split more than 10 million years ago. Insectivorous birds have been around all that time. Are we looking at a color pattern selected for by birds 10 million years ago and persisting in these two very different lineages to the present, or are we looking at a color pattern evolved repeatedly and independently? Given the presence of this color pattern in very different lineages of both pyrgines (*Phocides, Tarsosoctenus,* and others) and phyrrhopygines (*Myscelus, Parelbella, Elbella, Jemadia,* and others), it is likely that we are looking at a combination of phylogenetic inertia and independent evolutionary reinvention of the wheel.

We don't know for sure, but it is probably nothing but a coincidence that both *P. macleannani* and *P. nigrescens* caterpillars feed on the same myrtaceous food plants in the ACG, in the same ecosystems and at the same time. Fortunately for the caterpillar inventory project, the caterpillars are very different: *P. macleannani* is the classical black and orange ringed hairy pyrrhopygine, whereas *P. nigrescens* belongs to the large group of white, bird-dropping mimics (or as mimics of white, mold-covered dead caterpillars or pupae). The two can be found together on the same individual plant, though 1 percent of the field-collected *P. macleannani* caterpillars have been found enigmatically eating large myrtaceous-looking leaves of Melastomataceae that are beige on the underside, as in *Miconia trinervia* and *Miconia dolichopoda.*

ADULT VOUCHER: 04-SRNP-24065; JCM
CATERPILLAR VOUCHER: 05-SRNP-41519; JCM

93. *RHESCYNTIS HIPPODAMIA* – SATURNIIDAE

Although our page-sized image puts *Rhescyntis hippodamia* on equal footing with what you may think of as the usual size of a moth or butterfly, this arsenurine saturniid moth holds the record for the largest wing area of any reared ACG moth or butterfly. The as-yet-to-be-reared *Thysania agrippina* (Noctuidae) has the largest wingspan of any moth or butterfly. The front wings show yet another way to be a dead leaf in the ACG rain forest. Also, you can see the origin of its species epithet; there is a Roman horse hiding in there somewhere.

The size of a small dinner plate, *R. hippodamia* adults are occasional catches at rain-forest light traps from Mexico to Brazil (Lemaire 1980), but their caterpillars have long been a mystery. Surely large enough to be encountered by chance, they are in fact not encountered by chance. We knew this to be an elusive species after years of searching. Then, on 20 March 1995, on a rain-forest understory sapling of *Virola koschnyi,* a common tree in the Myrsticaceae (the nutmeg family), Lucia Ríos Castro, working out of Estación Pitilla, encountered what seemed to be an absolutely enormous, finger-sized, slug-like caterpillar, colored the same yellowish beige as the underside of the *Virola* leaves. Was it a huge lycaenid? No way. Was it a huge limacodid? Again, no way. Neither was possible, but ignoring the gigantism, that's the appearance the caterpillar held. It also had the sluggish behavior of both, but since there is no lycaenid or limacodid adult worthy of such dimensions, it was easy to know what it was not. Thirty suspense-filled days later, the answer was delivered in the form of an adult. Since this single encounter, no free-living caterpillar of *R. hippodamia* has been found within the ACG, despite intense search of this same forest and more than 700 individual caterpillar records from *V. koschnyi* and *Virola sebifera,* but all from plants one to three meters tall. We now know that foliage of *V. koschnyi* is an excellent food for *R. hippodamia,* having reared caterpillars from eggs obtained from a female found in the forest while in the act of mating.

94

Why has this caterpillar been field-collected only once, even with knowledge of a suitable food plant? The most likely scenario is that *R. hippodamia* is a moth of the crowns of *Virola* trees twenty to forty meters tall, not descending into the lower canopy to oviposit. The single, last instar that Lucia found on a low sapling probably fell from the crown as a younger caterpillar and found a suitable food plant at ground level, avoiding starvation. Alternatively, some addled or exhausted female might have placed an egg on that sapling.

Because of an accident of nature and the searching-collecting-rearing protocol, we learned what the caterpillar of *R. hippodamia* looks like and what it eats. We also learned how to rear it, should a fertile female become available. But overall we have only the slightest hint of the ecological, physiological, behavioral, and evolutionary biology of the species. In time, with constant vigilance and careful documentation, certain aspects of the natural history will be discovered. The process of integrating new information with previous knowledge, resulting in the eruption of new questions, is how the caterpillar inventory project advances.

ADULT VOUCHER: 98-SRNP-7679.03; JCM
CATERPILLAR VOUCHER: 95-SRNP-664; DHJ

94. *ROTHSCHILDIA ERYCINA* – SATURNIIDAE

An adult *Rothschildia erycina* is the official logo of Area de Conservación Guanacaste. This peculiar saturniid was not meant to be the logo. The larger, more common, and much better known *Rothschildia lebeau* (Janzen 1984b) was selected in the late 1980s to be the ACG logo. However, by accident a specimen of *R. erycina* was passed to the artist by bioilliterate administrators and thus became history. There are many members of the genus *Rothschildia* in the Neotropics, most of them looking like *R. lebeau* (Lemaire 1978), but *R. erycina* departs from this pattern by having the tips of the forewings greatly exaggerated, and the overall body and wings strongly reduced in size. While the selection for this change is long lost in evolutionary history, it is possible that it had to do with selection favoring smaller adults in a food-poor world or season.

Along with this change in adult morphology came a departure from the usual world of active moths, the dark of night. While female *R. erycina* do search for oviposition sites in the dark of night, and thus are caught at light traps, the males are active largely at dawn, and thus are only very rarely caught in light traps. This is the primary reason why this moth is so very rare in collections (most of the moths caught by light traps are males). The males fly at dawn because the females often do not start calling with pheromones until shortly before dawn, and continue to call well into full daylight (though not direct sunlight). It is a truly beautiful sight to see a half-dozen male *R. erycina* dancing in the air around a female in early morning light in the dry forest. There is a hint that in the rain forest, *R. erycina* females may call as early as 3 AM, with mating then occurring in full darkness, but in ACG dry forest, dawn appears to be the primary time of courtship.

Brightly ringed in red, whitish-green, and black, the caterpillars of *R. erycina* depart from the usual *Rothschildia* mode of cryptic green with nonurticating spines that mimic the spines of urticating saturniids and limacodids. Not only do the caterpillars bear the ringed colors of a coral snake and therefore may be deduced to be aposematic mimics, but insectivorous birds studiously ignore them—and it must be an intentional act of ignoring since the eggs are laid in clusters of forty to one hundred and a single tree may be festooned with the brightly colored caterpillars. When fed to nestling trogons (*Trogon elegans*), they caused no ill effects whatsoever. The caterpillars seem to have an internal compass that they use in the middle of the day. They walk in straight lines many tens of meters in length when they come down out of the tree to seek a place to spin a cocoon.

ADULT VOUCHER: 01-SRNP-10911; JCM
CATERPILLAR VOUCHER: 82-SRNP-804; DHJ

95

95. *GRETA MORGANE* – NYMPHALIDAE

No, the wings of *Greta morgane* are not glossy black, they are transparent. Over most of the wing surface, above and below, the scales have been reduced to almost non-existence and the wing is made of such clear cuticle that you can read very fine print undistorted through it.

You may think you have seen *G. morgane* buzzing, flitting, flapping, or sailing along a rain-forest or cloud-forest understory trail. Actually, what you saw was one of at least twenty-five

Costa Rican species of look-alikes (see images in DeVries 1987; also note, the *G. morgane* color pattern is almost entirely absent from ACG dry forest). This clearwing with a black, brown, and white-patterned margin is deceptively easy to recognize without thinking about what it really means. As caterpillars, ithomiine nymphalids, of which *G. morgane* is a prime example, eat the foliage of species of Solanaceae. They are probably correctly viewed as highly inedible owing to nasty alkaloids (nicotine, etc.) sequestered from the food plant. However, they are not merely aposematic in color and ostentatious behavior. The clear wings make them very cryptic when perched in the dark and dappled shady understory of the forest. There is evident value in being nonaposematic, so to speak, most certainly when perched and to some degree when being pursued. This implies that there is some predator that can in fact eat them, unless their huge Mullerian mimicry ring is an anachronism, a holdover from a time of other predators. This seems most unlikely, but it is very difficult to guess just which species might be the visually orienting predator that gives these butterflies higher fitness if they are more transparent. Certainly the bird community of the ACG forest understory is not seen chasing and hawking clear-winged ithomiines flitting, seemingly floating (and easily captured) through the undergrowth. Clearwings are among the most consistently encountered of all rain-forest understory butterflies. A hand-reared insectivorous owl that had never seen a butterfly rejected clearwinged (and other) ithomiines from the first encounter (Janzen and Pond 1976). It did not go through any taste-and-reject experience to bring on the avoidance response.

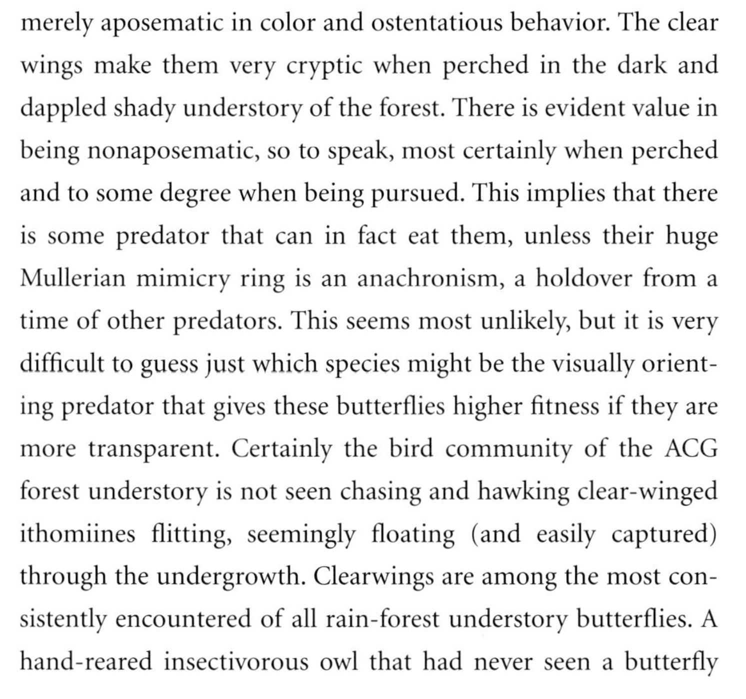

This implies to us that the kind of protection provided to the butterfly, and the members of its Mullerian mimicry ring, comes from genetic hard wiring in the predator to avoid such objects rather than a learned "tasting experience" as a younger bird. This constitutes instinctive avoidance. This same owl was offered other species of butterflies in this experiment and responded similarly to other aposematic patterns (see the species account for *Mechanitis polymnia*, #41).

ADULT VOUCHER: 04-SRNP-49510; JCM
CATERPILLAR VOUCHER: 98-SRNP-14934; DHJ

96. *MYDROMERA NOTOCHLORIS* – ARCTIIDAE

Anyone who sees the black, bright blue, and white ctenuchine arctiids buzzing around like big, vicious wasps in broad daylight in ACG rain forest, as does *Mydromera notochloris*, tends to think of them as Mullerian mimics. Indeed, they are mimics, as can be attested to by anyone who has been stung by one of the wasps. Because they are arctiids, which are also widely thought of as highly toxic to eat, it is assumed that the caterpillars feed on some type of noxious plant (e.g., Apocynaceae, Asclepiadaceae, Moraceae, Loganiaceae, Asteraceae) and either directly sequester toxic small molecules from these plants, or use them as building-block molecules to construct their own.

It was therefore startling to us to discover that the caterpillars of *M. notochloris* feed on Cyperaceae, the sedge family. Although tropical sedges may contain some moderately toxic defensive compounds, practically all plants do, and sedges are generally thought of as being about as toxic as grasses, which are usually protected from herbivores by more physical defenses, such as silicon crystals that wear down the mandibles of the chewer. The caterpillars of *M. notochloris* do resemble ctenuchine arctiids in their hairiness, which is interspersed with naked patches, but they do not radiate other warning signals. They tend to run when molested (making them doubly difficult to photograph), a trait that is often linked to being edible.

Adult *M. notochloris* have the coloration normally associated with deep rain-forest species, but the adults and caterpillars have been found throughout the lowlands of ACG, from the driest tip of the Santa Elena Peninsula to the edges of deep rain forest at Estación Caribe. Its presence in the driest parts of the ACG suggests that it too might be a dry-forest species, or simply cosmopolitan at low elevations. Alternatively, it could be a dry-forest species that has followed the sedge-filled pastures into the rain forest as they dry and insolate the habitat. Will the population of *M. notochloris* increase or decrease as the sedge-filled pastures give way to forest with a shaded understory? The current generation of elder biologists may never know, but some day when the ACG is in an advanced state of restoration, the rain forest will provide the answer.

ADULT VOUCHER: 04-SRNP-30836; JCM
CATERPILLAR VOUCHER: 04-SRNP-20879; DHJ

97. *CACOSTATIA SAPPHIRA* – ARCTIIDAE

Can you tell *Cacostatia sapphira* from *Belemnia trotschi* (#98) from *Mydromera notochloris* (#96) at a glance? In the static portraits it is easy to see differences in white versus orange bands on the forewing; dark brown, black, and deep blue background colors; and various degrees of iridescence on the abdomen. Now try it when they are buzzing around among the dappled shadows of their food plants one meter above the ground. These diurnal species of arctiids nonchalantly fly throughout the bright sunny day in front of birds that would have no problem snatching them from the air.

97

The caterpillar of *M. notochloris* eats sedges (Cyperaceae), and the caterpillar of *B. trotschi* eats the foliage of fig trees (Moraceae). The jury is definitely still out as to whether these species are Batesian or Mullerian mimics. If they are Mullerian mimics, then they must be manufacturing their own defensive chemicals. However, the caterpillar of *C. sapphira* feeds on the latex-rich leaves of

Stemmadenia and *Tabernaemontana* (Apocynaceae), and odds are that it sequesters some of the plants' alkaloids, reinforcing the aposematic appearance in the adult.

What is milky latex? The latex itself, a liquid rubber if you will, is probably not the toxin that leaves apocynaceous plants the last stalk standing in an overgrazed cattle or horse pasture. Rather, it is probably some nasty small molecule, such as an alkaloid or cardiac glycoside, present in the latex that is the plant defense against vertebrate herbivores. In turn, the successful insect herbivore acquires the toxins and uses them as a protective chemical against vertebrate predators.

A word of caution is appropriate for students of this and many other ostentatious ACG moths and butterflies. These moths may be gaining their protection not because the birds have suffered a quite unpleasant event after swallowing one, but because the birds are genetically hard-wired to avoid eating them. Credence to this idea comes not only from experiments testing bird reactions to the rings of the coral snake (Smith 1975, Janzen and Pond 1976), but also from the large number of species of moths that are small, black, and iridescent blue with oblique, contrasting bands on the forewings. These moths circulate in the daytime in ACG lowland forest. In particular, there are many other small black and white moths, notably in the genus *Desmia* (Crambidae). The crambids do not fly freely in sunlight but do launch freely into a blur of black and white buzzing flight when the foliage they are on is disturbed. Their readiness to launch and expose themselves, in contrast to the obviously cryptic and bark-colored or green moths, which tend to hang or sit tight, strongly suggests that such black and white moths are either aposematic (not likely, though possible) or part of a mimicry ring that includes *Cacostatia, Mydromera,* and perhaps even *Belemnia.* This being the case, then the extremely prevalent crambid mimics are much more abundant than their arctiid models. That observation does not fit well with classically defined mimic-model relationships, in which the model is deemed to be more abundant than the mimic. An alternative nominee for the model would be some species of social wasp. It makes sense, but we just do not know for sure.

ADULT VOUCHER: 04-SRNP-55538; JCM
CATERPILLAR VOUCHER: 03-SRNP-6236; DHJ

98. *BELEMNIA TROTSCHI* – ARCTIIDAE

This diurnal moth falls into the category of unbelievably gaudy. All that color is no accident of pigment physiology. *Belemnia trotschi* is undoubtedly saying something to the world, but what? The widespread belief among entomologists is that such bright colors are aposematic. This may well be the case, but there are caveats.

First, the known ACG food plant of the very cryptic caterpillar is *Brosimum guianense* (Moraceae). The fruits, seeds, and foliage of this large tree are edible to vertebrates, meaning the plant does not contain toxic chemicals. The inference is that if this arctiid moth is toxic to the touch, foul tasting, or gut wrenching (as we believe many arctiid moths to be), it is very likely manufacturing its own toxins.

Second, we need to ask if the colors are displayed ostentatiously, or does this moth in nature actually match some gaudy background, like a flower or ripe fruit? The short answer is that the adults do walk and fly with impunity in the full sun over foliage and flowers of many colors, but none of them are colors that match the moth. The moth is unambiguously displaying a warning signal.

Third, a gaudy arctiid may be a Batesian mimic, quite edible while looking much like another truly obnoxious species. In the case of *B. trotschi,* it is unique, not looking like anything else in any tropical habitat known to us, though one can never fully discard the possibility that at the time of the evolution of these colors, or in some other part of its range, there were a host of other look-alikes.

Fourth, it may be that these colors are simply a gaudy mess that transmits the signal, "I am one of those bright gaudy objects that you don't want to encounter close up." As such, it would be sharing a defense strategy with many tens of species of diurnal Arctiidae, Riodinidae, Noctuidae, Geometridae, and Nymphalidae, all brightly colored but each in its own fashion.

In short, we will never really know the function of these brilliant colors until someone conducts choice experiments with potential predators in its habitat, and even unpleasant (because the study may result in the death of the test subject) feeding experiments with naïve predators. Although there will be no single answer that works for all places at all times, at least the question can be addressed experimentally, since we now know how to rear *B. trotschi.*

ADULT VOUCHER: 04-SRNP-42383; JCM
CATERPILLAR VOUCHER: 04-SRNP-3927; DHJ

99. *MESOTAENIA BARNESI* – NYMPHALIDAE

Named for the personal valet of William Schaus, who a hundred years ago gave us the first and most intense look at Costa Rican Lepidoptera, this gaudy tropical butterfly, black with deep iridescent blue markings on the dorsal side, occupies a precarious niche in the ACG. *Mesotaenia barnesi* (previously known as *Perisama barnesi,* see DeVries 1987) is one of the two Mexico-Central America endemic members of a species-rich South American genus, all of which are high-elevation butterflies with an unknown natural history. They are highly prized by collectors, but that is not what threatens them. It is easy to visualize *Mesotaenia* as having emerged from an isolated species of

Callicore or *Diaethria* occurring in the high, cold ecological island of a mountaintop (e.g., Hall 2005). The danger of habitat loss is the issue where the threat exists. Global warming is moving the warm lowland climate regime up the slopes of the volcanoes (Pounds et al. 2006). As the air warms in the ACG, species that are mountaintop endemics, which are adapted to relatively colder environments, will be pushed out of their habitat with nowhere to escape. There is no alternate habitat for them to immigrate or crawl off to. Perhaps the tops of taller mountains will serve as potential, albeit ephemeral, refuge? That would be truly a temporary reprieve for *M. barnesi,* for there are no Costa Rican mountains high enough to escape the warming from the lowlands.

Whereas *Callicore* and *Diaethria* caterpillars are feeders on vines and treelets of Sapindaceae, *M. barnesi* has moved onto its sole ACG food plant, *Weinmannia wercklei,* a shrubby treelet at an elevation of 1,400 meters on Volcán Cacao and a species of the family Cunnoniaceae, inhabiting the upper elevations. It very likely feeds on species in the same family on the top of Volcán Orosí and the Rincón de la Vieja complex, but no caterpillar inventory has been conducted there. Not surprisingly, the spiny (but harmless) green caterpillar is similar to the spiny (but harmless) green and brown caterpillars of *Callicore* and *Diaethria.* Now knowing that a member of *Mesotaenia* feeds on Cunnoniaceae, it should be possible to fill the gap identified by DeVries when he commented that "There is absolutely nothing known about its hostplants or early stages . . . any biological observations should be published, particularly any indication of its larval hostplants."

ADULT VOUCHER: 03-SRNP-3892; JCM
CATERPILLAR VOUCHER: 02-SRNP-9397; DHJ

100. *PIERELLA HELVETIA* – NYMPHALIDAE

Today *Pierella helvetia* is placed in the large family Nymphalidae, but for the past century or more it would have been placed in the Satyridae, and now in the Satyriinae, a subfamily of Nymphalidae. In the lowland tropics, satyrids, as their name implies, are brown and soft-winged understory species with quick and often interrupted flight, very often having false eyes on the wings. Though many extra-tropical species have the same flight pattern, there are also many satyrids of meadows, field edges, and other more insolated circumstances.

100

Floating over the forest floor in the deeply shaded parts of the ACG, *P. helvetia* is one of the few large butterflies that can be reliably encountered on a nonrainy day by walking for hours on a rain-forest trail, slowly and observantly. Perched motionlessly among the fallen and dark rotting leaves, this butterfly is close to invisible with its wings closed. The bright red of the upper side of the hindwings is not displayed until the butterfly launches into fast, darting flight among the leafless plant bases sticking up out of the litter. Then it perches again and the red tracking patch abruptly disappears. Additionally, the large, false eyespot in the upper corner of the dorsum of the hindwing is matched by an even better-defined false eyespot on the underside. Somewhat cryptic at a distance but quite convincing close up, this pattern gives the butterfly a split-second head start in its flight from a startled potential predator.

Caterpillars of satyrids feed on monocots: grasses, sedges, palms, and even some primitive plants. None eat the leaves of broad-leafed plants, the angiosperms, and *P. helvetia* is no exception. Its caterpillar finely mottled brown and gray, feeds on the leaves of understory *Calathea* (Marantaceae) and *Heliconia* (Heliconiaceae). There is even one rearing record of a caterpillar collected from a sedge species (Cyperaceae). The caterpillar usually feeds at night, hiding motionless during the day low down on the stem or even in the litter at the base of the food plant. In the ACG, *P. helvetia* is the rain-forest *Pierella* whereas *Pierella luna,* similar but lacking in the red patch, occupies the interface between dry forest and rain forest, its caterpillar feeding on the same food plants and behaving in the same manner.

ADULT VOUCHER: 00-SRNP-4485; JCM
CATERPILLAR VOUCHER: 03-SRNP-30519; DHJ

Life History and the ACG Environment

Life history traits related to growth and survival are integral to understanding the natural history of biological diversity. In fact, the diversity of adaptations regarding physiology, behavior, and morphology, many of which are noted in our species accounts, reveals the variety of character states that have evolved in response to predictable and unpredictable aspects of the environment. To further illustrate some of these adaptations, we have selected a limited suite of species for a discussion about puddling, nectaring, life span, nonfeeders, winglessness, dormancy, and movement.

As butterflies attempt to evade their predators, the male in particular has duties to perform before mating. Aside from responding to pheromones, one of the activities tied to mating success is sipping moisture from wet soil or moist animal excrement, an activity called puddling. The behavior of puddling is, in part, a prenuptial activity and is not practiced by moths, but many species of butterflies engage in puddling. However, many species of moths will respond to sites of fermenting sap on trees, as do butterflies.

A tourist driving from park to park across Costa Rica or walking the muddy, wet ACG roads cannot help but notice the clouds of butterflies perched on muddy roadsides and stream edges. Most of these puddling species are brightly adorned with yellow and white patterns, though a few are darker and more cryptic, often overlooked. The explanation that they are simply drinking water turns out to be quite wrong. The first clue is that puddling butterflies are essentially all male. This is illustrated quite dramatically in the drawers containing mounted butterflies in large insect museums, where the male specimens of these species can outnumber the female specimens a hundredfold. The males were caught easily while puddling. The few females that were captured were found either while they were flying by or at a flower, or they were reared from a caterpillar. The same disproportion in the gender ratio often shows up in moth collections, but here the reason is that most moths that come to lights are males. It is the males that are most active at night because they are in search of females.

Why do only male butterflies puddle? The easy answer is that they are not in fact drinking for hydration, but rather passing large quantities of water through in order to sequester the minerals, apparently primarily sodium, which they add into the sperm packet that they deliver to females. The females mate frequently, evaluate the quality of the male based on the amount of sodium and probably other contents, and make a physiological decision to either "digest" the sperm packet or use the sperm to fertilize her eggs. The arriving male settles in to "drink" side by side with others, the same and different species, apparently using their presence to determine where the sodium (or other minerals?) is sufficiently abundant to be worth mining. If he is the first to arrive at a wet, muddy site, he walks or flies short distances between tries, sampling the mud until he finds the right place. What is the right place in the thousands of meters of wet riverbank or roadside? Commonly it is where someone or something has urinated or defecated, creating a local sodium deposit.

Puddling by butterflies is not the only butterfly activity that resembles eating and drinking, but is fantastically more complex. The nectar in flowers varies among species, and even among times of day, in the amounts and kinds of sugars and amino acids it contains. This can be as much a reason as to why different species of flowers are visited by different species of butterflies, birds, bats, bees, flies, and wasps as are their colors, times of day of opening, duration, sizes, and shapes. A butterfly trap baited with rotting fish or carnivore feces catches quite different species and sexes as the same trap in the same place baited with fermenting fruit. A moth perched on the back of a sleeping bird with its proboscis inserted between the bird's feathers could be picking up about any kind of imaginable liquid food, and the liquid itself may have been deposited there by the moth to moisten or partially liquefy yet some other kind of food. The same applies to a *Heliconius* butterfly thrusting its tongue deep into a curcurbit flower and bringing it out encrusted with pollen. The pollen is no simple contaminant. Some pollen may survive to pollinate a later flower, but the great bulk is soaked in regurgitated nectar from some other species of flower, the amino acids leached out into the regurgitate, and the regurgitate sucked up again as a nitrogen-enriched broth. Quite independently evolved, a similar trick is performed by skippers perched on a dry, white bird dropping on a leaf. Close examination will show that the skipper is defecating on the dry bird waste, which moistens it, and then its long tongue is curved under the body to slurp up the highly desirable and intentionally concocted soup.

The maximization of longevity seems to be a human trait. Actuarial tables, insurance, and retirement are issues of the human condition. In moths and butterflies, a typical life cycle lasts for one year (univoltine), and the amount of time any one of the life stages persists is the grist of adaptation to various environments and the evolution of insect life-history strategies (Miller 1993, 2004a, b). In some species the adult life stage, egg stage, or pupal stage may exist for several days or, in a different species, perhaps up to ten or eleven months.

We really do not know exactly how long adult moths and butterflies can and do live in nature, and we will not know until nanotechnology gives us a tracking technique that we can apply harmlessly to an adult. But there are many indirect indications of longevity, each layered with qualifiers, largely because we rear some adult butterflies and moths in captivity. Just as with mammals, there can be huge differences in the adult life span depending on time, place, generation, nutrition, and random luck.

There are many species of ACG moths, and perhaps a very few butterflies, that neither eat nor drink as adults, which naturally limits their life span. All of their nutrients and water were obtained as a feeding caterpillar and then carried through the pupal stage into the adult. Their physiological adult life span is generally five to ten days. Whether they suffer higher or lower per-day mortality from external causes than do species of feeding moths and butterflies is simply unknown, though it is easy to imagine that eliminating the search for and use of adult food could improve the chances of surviving any given twenty-four-hour period. Searching for food undoubtedly increases exposure to predators and the vicissitudes of the physical environment.

There are at least two species of ACG moths that give us more detailed information about being a nonfeeding adult. *Arsenura armida,* a large saturniid, and *Hylesia lineata,* a small saturniid (for details, see our previous volume, *100 Caterpillars*), are characterized by adults that lay all their eggs in one mass (*A. armida* rarely deposits two masses; Costa et al. 2004). Once the eggs are laid, the moth is genetically dead, because the female has no further chance of contributing directly to the gene pool. However, presumably selection has not been able to push her evolution to the extreme. Pursuing a single line of thought, one might think of natural selection as having created an organism that puts every last unit of resource into one more egg, so that death follows immediately after. But the females continue to live and fly for at least one and perhaps two nights after laying their entire egg load.

How can selection possibly generate such a machine? While an absolute answer remains the domain of future research, there is at least one reasonable natural history scenario. Many females will mate on the first night as an adult, sometimes immediately upon emergence from the pupa and often prior to her first flight. But it is quite possible that naturally occurring fluctuations in population density or within-population variation in the timing of emergence occasionally may lead to no male arriving the first or second or some later night. There is, therefore, natural selection for a certain number of days of reserve life for the adult, rather than putting all the resources into eggs in a female that must achieve mating the first night.

An exception to this scenario regarding resource allocation is the evolution of wingless females. In a select few species, the females are wingless. Having put nearly all of her nutritional resources into egg production (none partitioned to flight muscle), she is merely a bag of propagules. She has a short life span and cannot disperse. The job of dispersal is the responsibility of the young caterpillars and is accomplished by being hairy and producing strands of silk, allowing them to float in the atmosphere and disperse in a relatively nondirectional manner. Also, these species appear to possess a relatively high reproductive capacity in compensation for extremely high mortality.

While today there is an abundance of food plants for both *A. armida* and *H. lineata* in ACG secondary successional dry forest, it is quite easy to imagine old-growth dry forest where their oviposition food plants were very scarce. This in turn could lead to the female having to spend several nights in search of a single food plant on which to deposit the entire egg load. Again, as with delays in getting a mate, such a delay may easily select for the female to be carrying enough food and water reserves to survive more than the minimum of the two nights (one to mate, the next to oviposit) required for the adult stage of her life. There is actually no possible selection to turn off her biological machine once she has laid her eggs. She simply runs down, like a car burning out the last drop of gas in its gas tank. All of this helps to explain why females of both species taken in ultraviolet (UV) light traps almost invariably contain no eggs.

Feeding adults have a quite different and much more complicated lifestyle than nonfeeding adults. The act of searching for and feeding at food sources—be they sodium-rich mud, amino acid–rich rotting animals, or nectar- and amino acid–filled flowers—selects for a variety of morphology and behaviors that are basically missing from the nonfeeding adults. For example, both sexes of adult Sphingidae, the hummingbirds of the moth world, hover in front of flowers and put their long tongues down the deep corolla for nectar. Natural selection acts for both sexes to hover well, thus wing size and shape in both sexes is close to identical in most species. On the other hand, Saturniidae, nonfeeding adults all, evolve males and females with quite different appearances. The males are relatively pointy-winged jets, racing to find females, whereas the round-winged, heavy-bodied females are helicopters, adept at moving among the foliage of their ovipositional plants (Janzen 1984a).

The caterpillar antecedent to the feeding adult is relieved from the constraint of having to get every amount and kind of nutrient needed by the adult from the leaves it eats. It seems almost certain that were it necessary to do so, its development would take longer, leaving it more exposed to carnivores. Some species of food plant would be unacceptable, reducing the range of species of food plants eaten by the total array of species of caterpillars.

The act of seeking adult food, a resource that can be so very different in location, morphology, and chemistry relative to the caterpillar food, opens possibilities for the adult life stage to expand into very different habitats and ecosystems. The requirement of needing two distinct types of food to complete the life cycle removes the distributional barrier imposed by the geographical limits of the caterpillar food plant.

Finally, the variable seasonality and success at locating and capturing adult food means that a given individual of a given species may have dramatically different relationships with plants at different periods in its lifetime. During the long dry season, a powerful, long-lived, feeding adult butterfly, such as *Archaeoprepona demophon* (#79) or *Agrias amydon* (#76), may survive for many months on trivial amounts of sugar from sap-oozing tree wounds. Then, with the coming of the rains, it may abruptly use its stored reserves to lay many single eggs on individually sought food plants and die in a few weeks. A noctuid moth may feed for a long time as a caterpillar, store up substantial fat reserves, eclose after a short period as a pupa, and then spend nearly a year as an adult, waiting for the next flush of leaves on which to oviposit while subsisting on trivial amounts of sugar from occasional flowers. Through such tactics as these, species have evolved a very different breeding seasonality than if they had to sit for a year as a dormant pupa, camouflaged from a passing armadillo or mouse snuffling through the litter in search of juicy morsels.

Although the ACG sits in the tropical belt of the New World, parts of it are deciduous during the lengthy dry season. Moths and butterflies are compelled to adapt to this dry season, and they do so in a variety of ways. For any given species, the survival tactic generally focuses on a particular stage in the life cycle: egg, larval, pupal, or adult. Across all of the species, certain life stages seem to be involved in passing the dry season more than others.

Motoring through the ACG rain forest, lush and green and often cool and wet throughout much of the year, it does not cross your mind to think of how moths and butterflies pass the bad time of year. Bad time? What and when is that? When you drive from east to west across the ACG, dropping into the Pacific dry forest that lies in the rain shadow of the Cordillera Guanacaste for six months of each year (December through mid-May), you might ask yourself where all the butterflies have gone. A light trap set up in the dry season attracts only 1 percent of the numbers and species of moths that the same trap would attract in the May to December rainy season. Then you ask the same question about the moths. The dry season in the dry forest is characterized by blistering heat and 30 percent relative humidity during daylight hours. March to early May seems like the California Death Valley of the tropics. Where have the Lepidoptera gone and how do they survive these climatic conditions? As usual in the tropics, the answers are complex and subtle. There is no deadly seasonal freeze or the nearly constant dry climate of a desert.

The three major ACG ecosystems—dry forest, cloud forest, and rain forest—range widely and interdigitate in many complex ways. They have also been subjected to 400 years of intermittent assaults, including logging, ranching, farming, hunting, burning, ground water deviation, and most recently global climate change. The photo on page 226 shows the view from west to east at Cerro Ingles, across the serpentine anthropogenic barrens of the Peninsula Santa Elena toward the nearly cloudless Cordillera Guanacaste in eastern ACG. The low, gray, deciduous dry forest on the north-facing valley side is what covered these hills prior to European-style exploitation, and the extensive yellowish grass is native but originally occurred just on ridge crests and very steep, unstable slopes. With the elimination of anthropogenic fires by ACG management, the forest eventually returns but is very slow to occupy these slopes. When the rains come at the end of the dry season, this forest turns bright green and feeds thousands of species of moths and butterflies.

The contrast between the dry and rainy seasons is illustrated by the pair of photos on page 227 of the Sector Mundo Nuevo boundary zone between the ACG (on the right side of each image) and the long-degraded, brushy neighboring pastures filled with introduced African grasses (on the left side of each image). In both habitats in the dry season (left image), al-

Habitats within the ACG—Cerro Ingles. Volcán Orosí at 1,400 meters is on the left, Volcán Cacao at 1,500 meters is on the right.

most no caterpillars are present. In the rainy season (right image), however, both regions are rich in numbers and species of moth and butterfly caterpillars, yet with far more species in the ACG.

As diurnal, sun-loving creatures, butterflies are not active on foggy days. A search for butterflies under these conditions is nearly fruitless, but searching for moths at night using a UV light trap during foggy conditions is highly productive. As for caterpillars, they are indifferent to the fog. Climates such as that shown in the photo opposite are gradually diminishing, literally

Sector Mundo Nuevo in the dry season (left) and in the rainy season (right)

being burned off the ACG volcano tops with climate change, every year shrinking this ecosystem a little more.

The dry and rainy seasons are nicely contrasted by comparing a creek bed in March to the same site in October (see photos on page 228). This creek occurs in the secondary successional, eighty-year-old forest that is so widespread in western areas of the ACG such as Cafetal, Sector Santa Rosa. In the image on the left, almost all the trees are leafless for many months, and little to no standing water exists. Then the rains come. Notice the high water in the same creek bed from the same view in the image on the right, taken during the "full blast" period of the rainy season. In this specific case, the creek is rushing with the water it received as rainfall runoff from the western edge of Hurricane Mitch in October, 1998.

Habitats within the ACG—foggy day at the intermediate-elevation rain-forest edge, Estación San Gerardo, Sector San Cristobal

The options for how a butterfly or moth species passes the long dry season are many, and all are used to some degree. In contrast to northern cold winters, however, dormant eggs are an extreme rarity in the ACG. To date, only one species of moth, out of thousands of species of moths and butterflies, is known to use this method for survival. The female *Hylesia lineata* lays her packet of several hundred eggs on a twig and wraps them in

Dry-season creek in March (left) and during the wet season in October (right), Cafetal, Sector Santa Rosa.

a dense felt of barbed and silky abdominal hairs (Janzen 1984c). The egg mass (photo on page 10) then sits dormant for the six-month dry season. The tiny caterpillars hatch with the onset of the first rains in May. All other species of ACG dry-forest moths and butterflies lay eggs that are ephemeral jewels for one to two weeks in duration.

Dormant caterpillars passing the dry season are likewise a rarity. There are none known that become dormant to pass the season and re-initiate feeding with the next rains. A very few species, mostly moths in their cocoons or hesperiid caterpillars in their leaf nests, pass the dry season as dormant prepupae. This tactic is most unusual because normally the prepupal stage lasts only a few days, the period of time between cessation of caterpillar feeding and voiding of the gut. It is usually simply part of the developmental preparation for molting to the pupal stage.

More commonly, there are many dry-forest species that pass the dry season, and even as much as the second half of the rainy season, as dormant pupae in the litter, in cocoons, in chambers in the soil, or even as naked pupae silked to a leaf on evergreen plants or a twig. Two species of *Protographium* (#55, #56) use the latter method to remain dormant for many months. However, the normal dormancy period for butterfly pupae is one to four weeks.

The great bulk of ACG species of moths and butterflies survive the dry season or other times when reproduction is less than optimal through unusual behaviors as adults (Janzen 1987a). In the ACG dry forest, they may migrate to distant, wetter ecosytems (as far away as the eastern Costa Rican rainforest or the volcano-top cloud forests), or they may simply move to moister local habitats (such as shady ravines) and hide in a crevice. During this time, such adults are in a state of reproductive diapause, meaning that the physiology of reproduction is turned off.

Each of these methods has its natural history and interpretive complications. For example, since we first encountered many of the ACG moths and butterflies in the dry forest, it is seductive to think of dry forest as their home ecosystem. However, it is biologically fair to recognize that many may be rain-forest species that move into dry forest for a single generation. They may seek new and relatively toxin-free foliage or freedom from carnivores, but they will always return to rain forest for the rest of the year. Some species of Sphingidae, such as *Xylophanes porcus* (#29), are less choosy. They have caterpillars in both dry forest (in the first half of the rainy season) and rain forest (all year). This species was thought to migrate back and forth between the two habitats, but may actually be two sibling species, one a dry-forest breeder and the other a rain-forest breeder.

When the rains come in mid-May, butterflies are everywhere and moths come to light traps in droves. The sudden onset of such a high abundance and diversity of adults is quickly followed by a great pulse of caterpillars and a higher rate of reproduction in the carnivores that eat them. It has long been the general belief that newly appearing adults all eclosed through the stimulus of the rains. We now know that those adults are a great mix of migrants, recently emerged individuals, those who fly in locally from sheltered refuges and moist havens, and local adults that have been reproductively dormant but quite capable of avoiding predators and visiting adult food sources when available (Janzen 1987a, b; 1993).

As the caterpillar inventory gradually expanded from ACG dry forest to the cloud forests and mid-elevations of the volcanoes, and to the Caribbean slope rain forests, it encountered a less synchronously pulsed world of active adults and caterpillar abundance. Although the cross-community seasonal pulses within these habitats were diminished in bulk, the seasonality of individual species and families became more noticeable. But the phenology of single-species population cycles and irregularities in their presence in the rain forest create extraordinarily unpredictable circumstances. Just when a high density of a given species in one year allows a generalization as to what time of year that species will be present and in which life stage, the following year it displays a quite different pattern. There are clearly times in a given year when it is easy to note that caterpillars or adults of many rain-forest species are abundant or absent, but it is not at all certain that the pattern will repeat itself in the following year. It appears that the abundance of many species is a result of a diverse set of signals, cues, and stimuli, as well as survival serendipity.

What's in a Name?

The simple fact that we need a name for each organism on Earth has occupied modern taxonomists, scientists who classify plants and animals, for over 250 years, and the job is not done, not by a long shot. We need unique names to be precise in how we communicate with one another about a species, to delineate one species from the next, and equally important, to record our past, present, and hoped for future experiences with a species. Latin binomials, consisting of a genus and species epithet, were invented to facilitate this process. Common names are easier to pronounce but fraught with clumsiness and error. For example, imagine transmitting vital information about "the silvery blue butterfly that flies like crazy when you get near, if it is not drunk on fermented tree sap." This is not a useful way to tell someone in modern-day Mexico or someone in Colombia 100 years hence that you have observed *Archaeoprepona demophon* (#79). Lengthy descriptions are useless in a world where there are several dozen species of silvery blue striped *Archaeoprepona,* with as many as four species occurring in the same hectare.

Just imagine telling a story about the interactions of these four species: "the silvery blue stripe," "the slightly bronzy silvery blue stripe," "the wiggly black line silvery blue stripe," and "the yellow-beige with silvery blue stripe." Even if you are happy with one of these compound common names, add in the complexity of discovering, after 200 years of study, that "the silvery blue stripe butterfly" is in fact two species, the caterpillars of one eating the foliage of the treelet with glossy, elongate green leaves and the other the foliage of the shrub with yellow-green leaves with scalloped margins. With over 10,000 species of Lepidoptera occurring in the ACG alone, as well as thousands of other animals and plants, scientific names are a necessity.

Fortunately, a universally accepted system for nomenclature, the naming of species, exists. The rules of nomenclature define the boundaries for naming species, and the rules of systematics provide the means for justifying a change in a name and deciding how the species is defined.

When we began taking inventory of the caterpillars of the ACG in the dry forest of Sector Santa Rosa in 1978, we aimed to find and rear caterpillars as well as catch and identify adults. Having a library of accurately named adults saved time and money every time a mystery caterpillar was reared into an adult.

Matching the reared adult to a pinned adult in a library assured us that the species had a clean taxonomy and a name, instead of needing many years of research.

The taxonomic library required to name the species exists among the specimens in the collections of the world's natural history museums and in the heads of the world's expert taxonomists. The cumulative sum of this information is based on more than 200 years of literature on the taxonomy of moths and butterflies around the world.

To apply this knowledge, taxonomists need good specimens: properly labeled, properly prepared, and in series to portray variation. They need accurate natural history information, not just from the area collected, but from the entire region—in our case, Costa Rica. So our inventory project naturally turned into one of collecting and rearing caterpillars by day (some by night), and light trapping for adults by night. The product of these activities was a pile of boxes containing pinned specimens. The specimens were first hand-carried to Philadelphia and thence shipped to experts around the world. Later the specimens were deposited into the nascent Costa Rican national biodiversity inventory, created with the birth of the Instituto Nacional de Biodiversidad (INBio) in 1989.

Gradually the names for the thousands of species of moths from the light traps and the butterflies collected with a net began to accumulate. The process continues to this day in the ACG and will persist for many decades. Some individuals within a species are larger than others, lighter or darker, or more or less patterned. Factors that affect this variation in appearance may be variation in genes, food availability, and the rigors of survival.

The pinned and spread specimens are the basis of the taxonomic library being built for future generations. The specimens will be perused and compared by a taxonomic specialist on that group. When we get lucky, the specimens are identified to the species level.

An unknown noctuid may come by its name fairly easily, for example. We visited a museum and peered into drawer after drawer of moths, our unknown moth in hand. Finally, we matched the specimen to a previously identified moth. It was pronounced to be *Gonodonta pyrgo* (#74). Ed Todd, a taxonomist, had revised the genus *Gonodonta* a few decades ago and curated the *Gonodonta* collection in the National Museum of Natural History. Thanks to his work, our specimens were easily compared to a series showing variation. As it turned out, *G. pyrgo* had been described about two hundred years ago based on specimens ranging from Mexico to Brazil, thus encompassing our location and giving us yet more confidence we had made an accurate identification.

Not all identifications go that well. More frequently, all we can do is compare our specimen with a small series of adults collected from a couple of locations scattered about the New World tropics sometime over the past 100 years. With luck, an expert at the time of curation named these specimens and placed them in an accessible museum, such as the Smithsonian Institution in Washington, D.C. or The Natural History Museum in London, among many others.

Sometimes we search in vain for a name. A regretful taxonomic specialist may say: "Well, I have seen some of those somewhere, but cannot remember when, and anyway, that genus needs revising and I am currently working on a different group. All I can say is that it probably has a scientific name but it will be years before that name gets applied to your specimens—unless you go peruse the holotype specimens at the American Museum of Natural History in New York and happen to find the type."

On a positive note, we have on occasion hit the jackpot when a taxonomist, confronted with many well-documented specimens generated by both the ACG caterpillar inventory and the ACG adult inventory, decides to abandon other directions in their research and takes on the ACG species (e.g., Burns and Janzen 2005a, b), much to the chagrin of biodiversity biologists elsewhere awaiting identification of their material. We consider ourselves especially lucky when we find a taxonomist who has spent forty years studying the moth or butterfly family to which many of our specimens belong. The taxonomist Claude Lemaire, who passed away in 2004, sent us a species-level identification by return e-mail after seeing just the digital image. Not only is there no successor to his expertise, there is unlikely to be one.

Over the past twenty-eight years, we have accumulated names for about 4,000 species through the assistance of over 150 taxonomists around the world. You may note that a few of the moths and butterflies in this book are represented by interim names rather than the usual italicized genus-species names. Two examples are *Astraptes* INGCUP (Hebert et al. 2004; #9) and the female of *Porphyrogenes* BURNS01 (#46). Taking the latter as an example, the species is known only from specimens reared from ACG caterpillars. No adult has ever been seen in the wild. For the moment, the undescribed species is tagged with the interim, unofficial name of BURNS01. The species is solidly placed in the genus *Porphyrogenes,* and it is likely to remain there, even after the genus is revised by Bernard Hermier, a French schoolteacher and an expert on the several dozen Neotropical members of this genus. John Burns, the Smithsonian Institution's taxonomic specialist on the family Hesperiidae, to which *Porphyrogenes* unambiguously belongs, will eventually baptize this species in honor of Peter Wege, a major donor to the conservation of the Rincón rain forest in which it occurs (see http://janzen.sas.upenn.edu). Within a few years, we will have a formal, scientific name to replace *Porphyrogenes* BURNS01.

Gene sequencing and other developments in the past few years have brought yet another wave of excitement into the taxonomic process. We are employing DNA barcoding (e.g., Hebert et al. 2003; Hajibabaei et al. 2006; Smith et al. 2006) and integrating the information with traits of caterpillar morphology, food plants, and habitat affinities. These data are then used to diagnose taxa that may be new. It has long been known that some "species," as we see them morphologically in the museum drawer, are actually complexes of extremely similar species. A new fast-track method for flushing out these species emerged in 2003, marking the beginning of the process toward automated specimen identification. DNA barcoding is the process of reading a short and standard set of DNA base pairs in the mitochondrial gene COI and using that as the signature for a species. After we had DNA-barcoded hundreds of species of ACG moths and butterflies, it became evident that some of the species were made up of unexpected complexes of similar species. Often, but not always, we find that the previously unnoticed members of a complex segregate so that each has a unique natural history or morphological traits that correlate with the unique DNA signature, the barcode. Identifying these unique segregates is made easier because we know the specimen's food plant, its phenology, and its microgeographic location, among other information. Then we can identify members of that species, confirming their existence with hardly more than a barcode. For example, *Astraptes* YESENN (#10) was for 200 years an unknown entity hidden within the morphologically defined species *Astraptes fulgerator.* However, morphological data, natural history, and barcoding suggested that *A. fulgerator* is composed of ten separate species (see Hebert et al. 2004).

Barcoding all of the ACG moths and butterflies, a three-year process now underway, introduces what may seem almost unbearable levels of complexity. For example, at least five species that we feature in this book are now considered to be a complex of sibling species based on the evidence from barcoding studies being conducted as we write this chapter. Even the precision of a scientific name bestowed today is subject to review as a consequence of future research and changing technologies.

The Parataxonomists: Lepidoptera and Plant Biologists

Gathering, rearing, and processing the countless numbers of caterpillars and adults that make up the collection from the ACG is laborious and painstaking work. Much of this work is completed by local experts, the parataxonomists. The sequence of events that starts with the discovery of a caterpillar on a plant in the forest ends with the deposition of an adult moth or butterfly into a museum drawer. We know what we know today about the natural history of moths and butterflies of the ACG because for over twenty-eight years more than twenty dedicated field biologists—parataxonomists, or in the local language *gusaneros* and *gusaneras*—have dedicated themselves to the job of collecting, rearing, and recording data about *gusanos*—caterpillars. The *gusaneros* and *gusaneras*, all Costa Rican, work at the eight rearing stations located within the ACG.

Each parataxonomist is focused on the caterpillar inventory of his or her portion of the three major ACG ecosystems—dry forest, rain forest, cloud forest, and intergrades between these environments. The inventory work requires that each *gusanero* and *gusanera* is able to identify hundreds of species of plants and hundreds of species of Lepidoptera. Someone is at every station every day of the year. Working as a parataxonomist is no desk job. In addition to extensive taxonomic expertise, finding, processing, and recording caterpillars and adults requires the use of all-terrain vehicles, chain saws, laptop computers, database software, and digital cameras. The search for caterpillars is conducted by visually searching the foliage of plant species, sometimes with no particular goal in mind, but often seeking certain plants considered high priority for the ACG inventory. When a caterpillar is encountered, it is placed singly in a bag with the foliage of the caterpillar's food plant. If the species of food plant is unknown, then the parataxonomist collects additional samples of plant parts for a plant taxonomist. The location and date of the collection is recorded. After a few hours of field collecting, the bagged samples are taken to the rearing station, where they are assigned a unique code in the database and hung on a line. Parataxonomists check the bags containing the caterpillars every day to replenish foliage, note changes in the caterpillar as it molts, record dates of pupation

The parataxonomists of the Area de Conservación Guanacaste, August, 2005

and adult emergence, and observe any disease or parasitoid activity. On any given day at any given rearing station, there may be many hundreds of hanging bags with either live caterpillars or pupae.

The emergence of an adult moth or butterfly triggers another round of database entries with identification updates. The moths and butterflies are placed in a freezer for transport to the project headquarters at Santa Rosa. Here Daniel Janzen sorts through the envelopes, validates identifications, and pins select specimens. All of the pinned specimens are dried in an oven and transported to the University of Pennsylvania in Philadelphia, where they are either deposited into the Janzen-Hallwachs ACG-reared Lepidoptera Collection or sent to a select few museums or recognized experts in the taxonomy of certain groups.

Although we've given this general description of our inventory process, it does not begin to account for the daily challenges and unusual circumstances each parataxonomist might face. On a typical day in April, a time when the hot dry season is in full glory, *gusaneros* are stationed all over the ACG.

SANTA ROSA, DRY FOREST

Guillermo Pereira Espinoza is making a list of food plants to reprovision forty-eight rearing bags, mostly from evergreen species in riparian habitats in deciduous forest of the Santa Elena Peninsula. He is doing this while replacing a bearing in the right wheel of the project's sport utility vehicle in the 38°C heat of the Santa Rosa dry season. He is also anxious for the day that the new apprentices, *Lucia Vargas* and *Dionis Rivera,* have the experience and knowledge to direct their own activities.

Ana Ruth Franco Guadamúz is removing moths and butterflies dried in the oven overnight from the spreading boards. These moths and butterflies were mounted the night before from frozen adults that emerged back in December, 2005. She puts a label on each pin-mounted specimen before packing the dried specimens in a tight, overlapping, shingle-like fashion. Daniel Janzen and Winifred Hallwachs must carry as many specimens packed in small wooden boxes as permitted through customs on the flight from Liberia, Costa Rica, to Philadelphia, Pennsylvania.

Fredy Quesada Quesada is teaching two newly hired apprentice parataxonomists, *Hazel Cambronero* and *Sergio Rios,* how to take digital photographs of the newly mounted moths collected at UV light traps for the BioLep Project. The BioLep Project has as its goal the analysis of the DNA barcode for all species of Lepidoptera within the ACG. This could total more than 10,000 species. He is wishing for a cooling breeze to flush out the oven-like heat inside the Santa Rosa research building.

LOS ALMENDROS, DRY-FOREST—RAIN-FOREST INTERGRADE

Elieth Cantillano and *Roster Moraga Medina* are bumping over the road through the Del Oro orange plantations to their section of rain forest on the ACG boundary to search for caterpillars feeding on *Theobroma angustifolium,* a wild relative of cacao—chocolate. They are also looking at a list of food plants to bring back to the station for re-provisioning rearing bags.

Lucia Ríos Castro is at a table in the rearing barn leafing through a stack of flattened rearing bags she has just removed from the freezer. Each bag contains a butterfly that emerged during the past week. She is entering details into the database, writing a voucher code label, putting the label and the moth into a small plastic envelope, clipping it shut with a clothespin, and putting it back into the freezer for delivery to Santa Rosa next month.

PITILLA, INTERMEDIATE-ELEVATION RAIN FOREST

Petrona Ríos Castro is driving the quadricycle down the rough dirt road toward Santa Cecilia with a list of food plants to collect for re-provisioning bags. He is also making a mental list of the food to buy for the evening meal for four moth collectors and seventeen visiting donors from Canada, Japan, and Sweden.

Calixto Moraga, while bantering with *Manuel Rios,* an apprentice parataxonomist who finds more caterpillars than anyone else, is searching through a stack of several hundred small glass rearing bottles, trying to figure out why a reared parasitoid wasp arriving at Santa Rosa in February 2006 had a voucher code from January 2005. He determines it was because the voucher number 06-SRNP-30014 was accidentally given the label 05-SRNP-30014 when its parasitic wasp eclosed on 19 January 2006.

SAN GERARDO, INTERMEDIATE-ELEVATION RAIN FOREST

Osvaldo Espinoza Obando is driving the station's sport utility vehicle to La Cruz to get groceries and a thirty-gallon drum of diesel fuel. He is also transporting several thousand frozen moths and butterflies from the San Gerardo freezer, which was full to capacity, to the primary holding freezers at Santa Rosa.

Elda Araya Martinez is bent over in the grass at the edge of a rain forest pasture trying to find again the caterpillars of the huge butterfly *Caligo oedipus* exactly where she found them on 14 December 2005, marking the first time the caterpillar had ever been found anywhere and the first unambiguous record of the species for Costa Rica. The many other species of *Caligo* are not known to eat grasses anywhere. Their huge caterpillars instead eat large-leafed monocots such as palms, Heliconiaceae, Musaceae, Zingiberaceae, Marantaceae, and Costaceae.

Gloria Sihezar Araya is walking slowly through the adjacent rain-forest understory, wishing fervently that some of the food plants on her list occurred at higher density. All the time she is wondering when she might actually step on a fer-de-lance, a highly feared venomous snake, instead of being startled out of her wits by standing next to one. She is also hoping that *Anabelle Cordoba,* one of the apprentice parataxonomists, remembered to put the tube of parasitoid wasps preserved in alcohol in the refrigerator.

Carolina Cano Cano is walking slowly through the same rain-forest understory, wishing she could find another tree of *Ficus pertusa,* the source of the two caterpillars of *Kloneus babayaga* found in the area by Gloria the year before. These were the first records ever of the caterpillar and its food plant and had been an unnoticed species of sphingid moth in Costa Rica (see Miller et al. 2006 for its portrait). She is filling a large plastic bag with the foliage of specific plants for refilling rearing bags. She is wondering when *Yesenia Mendoza,* an apprentice parataxonomist, might arrive at the rearing barn to assist in finishing the day's work.

CARIBE, LOWLAND RAIN FOREST

José Manuel Peréz Fernández is fixing a flat tire on his motorbike and entering the date of the initiation of the prepupal phase for the caterpillars in the rearing bags just brought from the rearing barn to the nearby house.

Minor Carmona Bonilla is sitting on the concrete floor of the rearing barn with nineteen piles of different species of plants spread out in front of him, putting a large sprig of the correct leaves into each of forty-six bags. He is wondering what the adult butterfly looks like for 06-SRNP-40904, whose caterpillar was just discovered on *Welfia* palms. The Caribe parataxonomists never get to see the adults that are produced from their reared caterpillars because they ferry the pupae to San Gerardo, where there is electricity and freezers.

GONGORA, DRY-FOREST—RAIN-FOREST INTERGRADE

Dunia García García is taking each rearing bag off the overhead line, dumping out the old food, setting aside the bags with prepupal or dead caterpillars, and clipping the bags of the same species of food plants together in clusters on the clotheslines. At the same time, he is making a mental list of how many branches of which species of plants must be collected in the afternoon while searching for new caterpillars. He is waiting for Harry and Manuel to get back from Estación Cacao, high on the volcano overhead.

CACAO, CLOUD FOREST

Harry Ramiréz Castillo and *Manuel Pereira Espinoza,* after having bumped up the side of the volcano on the quadricycle to 800 meters elevation, are walking at 1,400 meters among the grass-like, short bamboo on the top of Volcán Cacao searching for hesperiid butterfly caterpillars. There is a vital need for more samples from the top, both because there are many undiscovered species at this site, and because this is documentation of an ecosystem that is on its way to extinction. They are wishing that the wind did not blow so hard and move the bamboo leaves so much. The caterpillars must be reared at the nearby Cacao station at 1,100 meters. They would die if taken to the lower and hotter elevations, a forboding consequence of impending changes in environmental conditions.

LA PERLA, DRY-FOREST—RAIN-FOREST INTERGRADE

José Cortés Hernández and *José Alberto Sánchez Chavarría* are negotiating a gully-washed, ancient logging road on the one-week-old quadricycle, both sweating under their motorcycle helmets in the 38°C dry-season sun. They finally arrive at a small pocket of evergreen treelets and search in vain for caterpillars. However, they both know that when the rains come a month from now, this habitat will be crawling with caterpillars.

Mariano Pereira Espinoza is sitting over a batch of inflated plastic bags, each with fresh green foliage collected an hour earlier along the river by the station. Each contains a caterpillar. He is looking up images in the project database, trying to match a riodinid caterpillar on *Inga* with an image in the database to get an interim field identification. He finally gives up and takes a digital photograph of the specimen.

BACK IN SANTA ROSA

All this time *Waldy Medina* has been in his office in Santa Rosa, trying to locate Camino Porvenir in the Rincón Rainforest sector so that he can put the record into the GIS database of collecting sites and assign its latitude, longitude, and elevation. He wishes that the *gusaneros* would not baptize their new collecting sites with names that contain long or confusable phrases and synonyms. Taking a break, he greets Daniel Janzen, who has arrived at Santa Rosa for a meeting about the new plan to fund an endowment for all of Costa Rica's system of conserved wildlands.

References

Acknowledgments

Numerical Species Index

Alphabetical Species Index

References

Burns, J. M. and Janzen, D. H. 1999. *Drephalys:* Division of this showy neotropical genus, plus a new species and the immatures and food plants of two species from Costa Rican dry forest (Hesperiidae: Pyrginae). *Journal of the Lepidopterists' Society* 53: 77–89.

——— 2001. Biodiversity of pyrrhopygine skipper butterflies (Hesperiidae) in the Area de Conservación Guanacaste, Costa Rica. *Journal of the Lepidopterists' Society* 55: 15–43.

——— 2005a. Pan-Neotropical genus *Venada* (Hesperiidae: Pyrginae) is not monotypic: Four new species occur on one volcano in the Area de Conservación Guanacaste Costa Rica. *Journal of the Lepidopterists' Society* 59: 19–34.

——— 2005b. What's in a name? Lepidoptera: Hesperiidae: Pyrginae: *Telemiades* Hubner 1819 [*Pyrdalus* Mabille 1903]: New combinations *Telemiades corbulo* (Stoll) and *Telemiades oiclus* (Mabille)—and more. *Proceedings of the Entomological Society of Washington* 107: 770–781.

Costa, J. T., Fitzgerald, T. D., and Janzen, D. H. 2004. Trail-following behavior and natural history of the social caterpillar of *Arsenura armida* in Costa Rica (Lepidoptera: Saturniidae: Arsenurinae). *Tropical Lepidoptera* 12: 17–23.

Covell, C. V. 1984. *A Field Guide to the Moths of Eastern North America.* Boston: Houghton Mifflin.

DeVries, P. J. 1987. *The Butterflies of Costa Rica and Their Natural History: Papilionidae, Pieridae, Nymphalidae.* Princeton: Princeton University Press.

——— 1997. *The Butterflies of Costa Rica and Their Natural History. Riodinidae.* Princeton: Princeton University Press.

Druce, H. 1900. Descriptions of some new genera and species of Heterocera from tropical South America. *Annals and Magazine of Natural History* 7: 507–527.

Gámez Lobo, R. 1991. Biodiversity conservation through facilitation of its sustainable use: Costa Rica's National Biodiversity Institute. *Trends in Ecology and Evolution* 6: 377–378.

——— 1999. *De biodiversidad, gentes y utopías: Reflexiones en los 10 años del INBio.* Santo Domingo de Heredia, Costa Rica: Instituto Nacional de Biodiversidad.

Hajibabaei, M., Janzen, D. H., Burns, J. M., Hallwachs, W., and Hebert, P. D. N. 2006. DNA barcodes distinguish species of tropical Lepidoptera. *Proceedings of the National Academy of Sciences* 103: 968–971.

Hall, J. P. W. 2005. Montane speciation patterns in *Ithomiola* butterflies (Lepidoptera: Riodinidae): Are they consistently moving up in the world? *Proceedings of the Royal Society of London B* 272: 2457–2466.

Hebert, P. D. N., Cywinska, A., Ball, S. L., and deWaard, J. R. 2003. Biological identifications through DNA barcodes. *Proceedings of the Royal Society of London B Biological Science* 270: 313–321.

Hebert, P. D. N., Penton, E. H., Burns, J. M., Janzen, D. H., and Hallwachs, W. 2004. Ten species in one: DNA barcoding reveals cryptic species in the neotropical skipper butterfly *Astraptes fulgerator. Proceedings of the National Academy of Sciences* 101: 14812–14817.

Hunt, J. H., Brodie, R. J., Carithers, T. P., Goldstein, P. Z., and Janzen, D. H. 1999. Dry season migration by Costa Rican lowland paper wasps to high elevation cold dormancy sites. *Biotropica* 31: 192–196.

Janzen, D. H. 1967. Interaction of the bull's-horn acacia (*Acacia cornigera* L.) with an ant inhabitant (*Pseudomyrmex ferruginea* F. Smith) in eastern Mexico. *University of Kansas Science Bulletin* 47: 315–558.

——— 1969. Allelopathy by myrmecophytes: The ant *Azteca* as an allelopathic agent of *Cecropia. Ecology* 50: 146–153.

——— 1977. Why fruits rot, seeds mold, and meat spoils. *The American Naturalist* 111: 691–713.

——— 1981. Patterns of herbivory in a tropical deciduous forest. *Biotropica* 13: 271–282.

——— 1983. *Erblichia odorata* Seem. (Turneraceae) is a larval host plant of *Eueides procula vulgiformis* (Nymphalidae: Heliconiini) in Santa Rosa National Park, Costa Rica. *Journal of the Lepidopterists' Society* 37: 70–77.

——— 1984a. Two ways to be a tropical big moth: Santa Rosa saturniids and sphingids. *Oxford Surveys in Evolutionary Biology* 1: 85–140.

——— 1984b. Weather-related color polymorphism of *Rothschildia lebeau* (Saturniidae). *Bulletin of the Entomological Society of America* 30: 16–20.

——— 1984c. Natural history of *Hylesia lineata* (Saturniidae: Hemileucinae) in Santa Rosa National Park, Costa Rica. *Journal of the Kansas Entomological Society* 57: 490–514.

——— 1985. On ecological fitting. *Oikos* 45: 308–310.

——— 1987a. How moths pass the dry season in a Costa Rican dry forest. *Insect Science and Its Application* 8: 489–500.

——— 1987b. When, and when not to leave. *Oikos* 49: 241–243.

——— 1988a. The migrant moths of Guanacaste. *Orion Nature Quarterly* 7: 38–41.

——— 1988b. Ecological characterization of a Costa Rican dry forest caterpillar fauna. *Biotropica* 20: 120–135.

——— 1993. Caterpillar seasonality in a Costa Rican dry forest. In *Caterpillars: Ecological and evolutionary constraints on foraging,* N. E. Stamp and T. M. Casey, eds. New York: Chapman and Hall, pp. 448–477.

——— 2000. Costa Rica's Area de Conservación Guanacaste: A long march to survival through non-damaging biodevelopment. *Biodiversity* 1: 7–20.

——— 2003. How polyphagous are Costa Rican dry-forest saturniid caterpillars? In *Arthropods of Tropical Forests: Spatio-temporal Dynamics and Resource Use in the Canopy.* Y. Basset, V. Novotny, S. E. Miller, and R. L. Kitching, eds. Cambridge: Cambridge University Press, pp. 369–379.

——— 2004. Setting up tropical biodiversity for conservation through non-damaging use: Participation by parataxonomists. *Journal of Applied Ecology* 41: 181–187.

Janzen, D. H., Hajibabaei, M., Burns, J. M., Hallwachs, W., Remigio, E., and Hebert, P. D. N. 2005. Wedding biodiversity inventory of a large and complex Lepidoptera fauna with DNA barcoding. *Philosophical Transactions of the Royal Society B* 360: 1835–1846.

Janzen, D. H. and Martin, P. S. 1982. Neotropical anachronisms: The fruits the gomphotheres ate. *Science* 215: 19–27.

Janzen, D. H. and Pond, C. M. 1976. Food and feeding behavior of a captive Costa Rican least pigmy owl. *Brenesia* 9: 71–80.

Janzen, D. H., Sharkey, M. J., and Burns, J. M. 1998. Parasitization biology of a new species of Braconidae (Hymenoptera) feeding on larvae of Costa Rican dry forest skippers (Lepidoptera: Hesperiidae: Pyrginae). *Tropical Lepidoptera* 9(suppl.): 33–41.

Kalko, M. and Kalko, E. K. V. 2006. Gleaning bats as underestimated predators of herbivorous insects: Diet of *Micronycteris microtis* (Phyllostomatidae) in Panama. *Journal of Tropical Ecology* 22: 1–10.

Kitching, I. J. and Cadiou, J. M. 2000. Hawkmoths of the world: An annotated and illustrated revisionary checklist (Lepidoptera: Sphingidae). Ithaca: Cornell University Press.

Lemaire, C. 1978. The Attacidae of America (= Saturniidae). Attacinae. Published by C. Lemaire, 42 Boulevard Victor Hugo, F. 92200, Neuilly-sur-Seine, France.

——— 1980. The Attacidae of America (= Saturniidae). Arsenurinae. Published by C. Lemaire, 42 Boulevard Victor Hugo, F. 92200, Neuilly-sur-Seine, France.

——— 1988. *The Saturniidae of America. Ceratocampinae.* San Jose, Costa Rica: Museo Nacional de Costa Rica.

——— 2002. *The Saturniidae of America: Part 4, Hemileucinae.* Keltern, Germany: Goecke & Evers.

Marquis, R. J. 1984. Natural history of a tropical daytime-flying saturniid: *Automeris phrynon* Druce (Lepidoptera: Saturniidae: Hemileucinae). *Journal of the Kansas Entomological Society* 57: 529–531.

Miller, J. C. 1993. Insect natural history, multi-species interactions and biodiversity in ecosystems. *Biodiversity and Conservation* 2: 233–241.

——— 1995. Caterpillars of Pacific Northwest Forests and Woodlands. USDA, USFS, FHM-NC-06–95.

——— 2004a. Insect life history strategies: Development and growth. In *The Encyclopedia of Plant and Crop Science,* R. M. Goodman, ed. New York: Marcel Dekker, pp. 598–600.

——— 2004b. Insect life history strategies: Reproduction and survival. In *The Encyclopedia of Plant and Crop Science,* R. M. Goodman, ed. New York: Marcel Dekker, pp. 601–604.

Miller, J. C. and Hammond, P. C. 2000. Macromoths of Northwest Forests and Woodlands. USDA, USFS, FTET 98–18.

——— 2003. Caterpillars and Adult Lepidoptera of Northwest Forests and Woodlands. USDA, USFS, FHTET-2003–03.

Miller, J. C., Hammond, P. C., and Ross, D. N. R. 2003. Distribution and functional roles of rare and uncommon moths (Lepidoptera: Noctuidae: Plusiinae) across a coniferous forest landscape. *Annals of the Entomological Society of America* 96: 847–855.

Miller, J. C. and Hanson, P. E. 1989. Laboratory feeding test on the development of gypsy moth larvae with reference to plant taxa and allelochemicals. Oregon State University Experiment Station Bulletin No. 674.

Miller, J. C., Janzen, D. H., and Hallwachs, W. 2006. 100 Caterpillars: Portraits from the Tropical Forests of Costa Rica. Cambridge, MA: Harvard University Press.

Miller, J. S. 1996. Phylogeny of the Neotropical moth tribe Josiini (Notodontidae: Dioptinae): A hidden case of Müllerian mimicry. *Zoological Journal of the Linnaean Society* 118: 1–45.

Pounds, J. A., Bustamante, M. R., Coloma, L. A., Consuegra, J. A., Fogden, M. P. L., Foster, P. N., La Marca, E., Masters, K. L., Merino-Viteri, A., Puschendorf, R., Ron, S. R., Sanchez-Azofeifa, G. A., Still, C. J., and Young, B. E. 2006. Widespread amphibian extinctions from epidemic disease driven by global warming. *Nature* 439: 161–167.

Ross, G. N. 2005. Survey of the butterflies of the Wah'kon-tah Prairie, Missouri. *Holarctic Lepidoptera* 8: 1–30.

Ruszczyk, A., Motta, P. C., Barros, R. L., and Araujo, A. M. 2004. Ecological correlates of polyphenism and gregarious roosting in the grass yellow butterfly *Eurema elathea* (Pieridae). *Brazilian Journal of Biology* 64: 151–164.

Schaus, W. 1910. New species of Heterocera from Costa Rica. *Annals and Magazine of Natural History* 86: 189–211.

Shapiro, A. M. 1976. Seasonal polyphenism. *Evolutionary Biology* 9: 259–333.

Smith, M. A., Woodley, N. E., Janzen, D. H., Hallwachs, W., and Hebert, P. D. N. 2006. DNA barcodes reveal cryptic host-specificity within the presumed polyphagous members of a genus of parasitoid flies (Diptera: Tachinidae). *Proceedings of the National Academy of Sciences* 103: 3657–3662.

Smith, S. 1975. Innate recognition of coral snake pattern by a possible avian predator. *Science* 187: 759–760.

Tuskes, P. M., Tuttle, J. P., and Collins, M. M. 1996. *The Wild Silk Moths of North America.* Ithaca: Cornell University Press.

Vaglia, T. and Haxaire, J. 2003. Description d'un nouveau Sphingidae du Costa Rica *Xylophanes letiranti* (Lepidoptera: Sphingidae). *Lambillionea* 103: 287–290.

Willmott, K. R. 2003. The genus *Adelpha:* Its systematics, biology and biogeography (Lepidoptera: Nymphalidae: Limenitidini). Gainesville, FL: Scientific Publishers.

Wolda, H. and Foster, R. 1978. *Zunacetha annulata* (Lepidoptera: Dioptidae), an outbreak insect in a Neotropical forest. *Geo-Eco-Trop* 2: 443–454.

Yack, J. E., Otero, L. D., Dawson, J. W., Surlykke, A., and Fullard, J. H. 2000. Sound production and hearing in the blue cracker butterfly *Hamadryas feronia* from Venezuela. *Journal of Experimental Biology* 203: 3689–3702.

Acknowledgments

FROM JEFFREY MILLER:

The work that went into creating the images and text for this book was in large part conducted simultaneously with the production efforts for *100 Caterpillars: Portraits from the Tropical Forests of Costa Rica,* published in 2006 by Harvard University Press and authored by myself, Dan, and Winnie. First and foremost, however, I would like to thank the team of parataxonomists that work in the ACG. The *gusaneros* and *gusaneras* have been very generous with their assistance during my visits, helping find gorgeous, huge, and ostentatious caterpillars to photograph among the hundreds of rearing bags hanging on the lines. Also generous in her time and assistance was Jean Miller, my wife and field assistant. She has been an integral part of the team from start to finish. She has endured long days in the tropical environment, going places and doing things that appall the staff of our local travel clinic. Her keen sense of record-keeping and eye for detail during editing have contributed in countless ways. William C. Krueger, head of my recently adopted Department of Rangeland Ecology and Management, has given me unconditional support to continue with entomological studies while my career is in transition from the context of the now-departed Department of Entomology. Oregon State University awarded me a grant (OSU 2003–2004 General Research Funds Grant Program) covering travel and expenses. For this I am extremely grateful—the financial and administrative support was a substantial morale booster. Joe Scheer of Alfred University hosted me as an artist-in-residence and provided unlimited use of his time and facilities.

His book, entitled *Night Visions,* his friendship, and his professional critique of my work have inspired and allowed me to become a better photographer. John Burns, our skipper expert at the Smithsonian Institution, kindly hosted me the day I took photographs of adults. Brent Helliker, a recently appointed assistant professor at the University of Pennsylvania, kindly allowed me to turn his remodeled lab into a photo studio for the four days I visited Philadelphia to photograph reared adults housed in the ACG collection. I would like to thank Harvard University Press, in particular Ann Downer-Hazell and Kate Brick. Ann was patient beyond belief, trusting that we could produce these images and the associated text without too much delay. Kate edited the manuscript with great care and skill, keeping what we wrote in our own style but more to the point. Lastly, Dan and Winnie not only most graciously invited me into their world of tropical caterpillars, but they helped make arrangements for lodging, provided transportation, and taught me all sorts of matters scientific, sociological, and political in a way that only they could do, in the first person, past, present, and future. What a grand experience this has been and I thank everyone who has endured my time crunches and over-exuberance throughout.

From Daniel Janzen and Winifred Hallwachs:

How do we know so much about the moths, butterflies, and caterpillars of the ACG? Because so many Costa Ricans have invested the best parts of their adult lives pretending to be birds and monkeys—and scorpions, ants, spiders, wasps, frogs, bats, mice, and all the other things that search for and eat caterpillars. They bring back the caterpillars with their food plants, babysit them through to adults, enter them into the database, and photograph them. These are the project parataxonomists, or *gusaneros* and *gusaneras,* with about 200 person-years of field work among them since 1988. They are Carolina Cano Cano, Elieth Cantillano Espinoza, Osvaldo Espinoza Obando, Ana Ruth Franco Guadamúz, Roster Moraga Medina, Guillermo Pereira Espinoza, Manuel Pereira Espinoza, Mariano Pereira Espinoza, Fredy Quesada Quesada, Harry Ramiréz Castillo, Lucia Ríos Castro, Gloria Sihezar Araya, Elda Araya Martinez, Petrona Ríos Castro,

Dunia García García, José Alberto Sánchez Chavarría, José Cortés Hernández, Minor Carmona Bonilla, Calixto Moraga, Manuel Ríos, Anabelle Cordoba, Waldy Medina, Lucia Vargas, Johan Vargas, Sergio Rhos, Dionis Rivera, and José Manuel Peréz Fernández. They, and their collection localities, have been guided throughout by the maps and GPS of Waldy Medina, the ACG and project Costa Rican GIS specialist. All of them have been assisted by a great number of volunteers ranging from members of the same family to visitors from many faraway countries.

We also know a great deal about the moths and butterflies of the ACG because hundreds of Costa Ricans from all social points of life have tolerated, facilitated, encouraged, helped, financed, stimulated, and absorbed this effort to inventory—to set up for all of society to do whatever they wish with them—the caterpillars, and now the adult moths and butterflies, of the ACG. We also know because U.S. tax dollars, filtered through the IRS, Congress, and the U.S. National Science Foundation (grants 8307887, 8610149, 9024700, 9306296, 9400829, 9705072, 0072730 and 0515699 to DHJ from Biotic Surveys and Inventories) have paid the lion's share of the bill. It is a fair estimate that the government of Costa Rica, through the Ministerio de Recursos Naturales y Energia (MINAE), the Sistema Nacional de Areas de Conservación (SINAC), the Museo Nacional de Costa Rica, and the ACG itself have contributed as much financial support in kind as the international community has contributed in dollars. While all of the ACG staff has been invariably supportive and facilitatory, we wish to thank specifically Roger Blanco, Maria Marta Chavarria, Julio Diaz, Luis Fernando Garita, Jose Jaramillo, Felipe Chavarria, Luz Maria Romero, Magda Rodriguez, Jose Antonio Salazar, Randall Garcia, Johnny Rosales, Guisselle Mendez, and Sigifredo Marin for a multitude of acts supporting the inventory. Ranging more widely throughout Costa Rica, we remember and gratefully acknowledge much explicit support for the inventory from Rodrigo Gamez, Luis Diego Gomez, Alvaro Umaña, Rene Castro, Isidro Chacon, Jesus Ugalde, Angel Solis, Jorge Corrales, Bernardo Espinoza, Jorge Jimenez, Jenny Phillips, Raul Solorzano, Mario Boza, and Alvaro Ugalde.

Most of all we know a great deal about the ACG caterpillars and their adults because nearly 150 insect and plant taxonomists all over the world have worked long, inconvenient, sweaty, back-aching, dusty, dull, exhilarating hours to help us identify the adult moths and butterflies, their parasitoids, and their food plants, of this caterpillar and adult Lepidoptera inventory. Special thanks are due to particular individuals who were key in the identification of the particular moths and butterflies that we display: John Burns, Claude Lemaire (deceased), Jean-Marie Cadiou, Ian Kitching, Dick Vane-Wright, Keith Willmott, Jason Hall, Don Harvey, Phil DeVries, Jim Miller, Jack Franclemont (deceased), John Rawlins, Isidro Chacon, Bernardo Espinoza, Bob Poole, Vitor Becker, Mike Pogue, Scott Miller, Marc Epstein, William Schaus (deceased), Alan Hayes (deceased), Paul Hebert, Mehrdad Hajibabaei, Alex Smith, John Wilson, Stephanie Kirk, Tanya Dapkey, Ronald Zuciga, William Burger, Gerrit Davidse, and Bernard D'Abrera. The parasitoid discussions would have been impossible without the identifications provided by Monty Wood, Norm Woodley, Ian Gauld, Rodolfo Zuñiga, Mike Schauff, Mike Sharkey, Carlos Sarmiento, Jim Whitfield, Alejandro Valerio, Josephine Rodriguez, Andy Deans, Won-Young Choi, Scott Shaw, and Nina Zitani. The food plants for the caterpillars displayed here would be just so much green salad without the identification labor freely offered by Nelson Zamora, Roberto Espinoza, Adrian Guadamuz, Maria Marta Chavarria, Luis Diego Gomez, Ron Leisner, Barry Hammel, Mike Grayum, Francisco Morales, Alexander Rodriguez, Luis Poveda, Jose Gonzales, Al Gentry (deceased), Jose Gomez-Laurito, William Haber, and Quirico Jimenez.

Numerical Species List

1. *Moresa valkeri*—Notodontidae
2. *Trosia* JANZEN01—Megalopygidae
3. *Automeris belti*—Saturniidae
4. *Automeris io*—Saturniidae
5. *Oryba achemenides*—Sphingidae
6. *Oryba kadeni*—Sphingidae
7. *Fountainea eurypyle*—Nymphalidae
8. *Siderone marthesia*—Nymphalidae
9. *Astraptes* INGCUP—Hesperiidae
10. *Astraptes* YESENN—Hesperiidae
11. *Chrysoplectrum pervivax*—Hesperiidae
12. *Neoxeniades molion*—Hesperiidae
13. *Eueides procula*—Nymphalidae
14. *Tithorea tarricina*—Nymphalidae
15. *Xylophanes tyndarus*—Sphingidae
16. *Adhemarius ypsilon*—Sphingidae
17. *Parides iphidamas*—Papilionidae
18. *Mimoides euryleon*—Papilionidae
19. *Bardaxima perses*—Notodontidae
20. *Calledema plusia*—Notodontidae
21. *Zerene cesonia*—Pieridae
22. *Epia muscosa*—Apatelodidae
23. *Cosmosoma cingulatum*—Arctiidae
24. *Cosmosoma teuthras*—Arctiidae
25. *Historis odius*—Nymphalidae
26. *Anaea aidea*—Nymphalidae
27. *Dysschema viuda*—Arctiidae
28. *Danaus plexippus*—Nymphalidae
29. *Xylophanes porcus*—Sphingidae
30. *Xylophanes crotonis*—Sphingidae
31. *Siproeta epaphus*—Nymphalidae
32. *Adelpha basiloides*—Nymphalidae
33. *Morpho theseus*—Nymphalidae
34. *Morpho amathonte*—Nymphalidae
35. *Opsiphanes cassina*—Nymphalidae
36. *Caligo atreus*—Nymphalidae
37. *Pentina flammans*—Thyrididae
38. *Othorene verana*—Saturniidae
39. *Memphis mora*—Nymphalidae
40. *Memphis proserpina*—Nymphalidae
41. *Mechanitis polymnia*—Nymphalidae
42. *Consul fabius*—Nymphalidae

43. *Arsenura drucei*—Saturniidae
44. *Mimallo amilia*—Mimallonidae
45. *Ferenta castula*—Noctuidae
46. *Porphyrogenes* BURNS01—Hesperiidae
47. *Oxytenis modestia*—Oxytenidae
48. *Copaxa curvilinea*—Saturniidae
49. *Myscelia pattenia*—Nymphalidae
50. *Hamadryas amphinome*—Nymphalidae
51. *Zale peruncta*—Noctuidae
52. *Letis mycerina*—Noctuidae
53. *Hypercompe icasia*—Arctiidae
54. *Neonerita dorsipuncta*—Arctiidae
55. *Protographium philolaus*—Papilionidae
56. *Protographium marchandi*—Papilionidae
57. *Kloneus babayaga*—Sphingidae
58. *Madoryx plutonius*—Sphingidae
59. *Oxynetra hopfferi*—Hesperiidae
60. *Creonpyge creon*—Hesperiidae
61. *Bungalotis astylos*—Hesperiidae
62. *Bungalotis diophorus*—Hesperiidae
63. *Sarsina purpurascens*—Lymantriidae
64. *Rejectaria atrax*—Noctuidae
65. *Erbessa salvini*—Notodontidae
66. *Zunacetha annulata*—Notodontidae
67. *Telemiades chrysorrhoea*—Hesperiidae
68. *Proteides mercurius*—Hesperiidae
69. *Zaretis ellops*—Nymphalidae
70. *Zaretis itys*—Nymphalidae
71. *Entheus matho*—Hesperiidae
72. *Dismorphia amphiona*—Pieridae
73. *Strophocerus thermesia*—Notodontidae
74. *Gonodonta pyrgo*—Noctuidae
75. *Heraclides astyalus*—Papilionidae
76. *Agrias amydon*—Nymphalidae
77. *Urbanus belli*—Hesperiidae
78. *Astraptes talus*—Hesperiidae
79. *Archaeoprepona demophon*—Nymphalidae
80. *Prepona laertes*—Nymphalidae
81. *Aphrissa statira*—Pieridae
82. *Phoebis philea*—Pieridae
83. *Aellopos ceculus*—Sphingidae
84. *Enyo ocypete*—Sphingidae
85. *Automeris metzli*—Saturniidae
86. *Automeris phrynon*—Saturniidae
87. *Cerura dandon*—Notodontidae
88. *Cerura rarata*—Notodontidae
89. *Azeta rhodogaster*—Noctuidae
90. *Pseudodirphia menander*—Saturniidae
91. *Phocides nigrescens*—Hesperiidae
92. *Parelbella macleannani*—Hesperiidae
93. *Rhescyntis hippodamia*—Saturniidae
94. *Rothschildia erycina*—Saturniidae
95. *Greta morgane*—Nymphalidae
96. *Mydromera notochloris*—Arctiidae
97. *Cacostatia sapphira*—Arctiidae
98. *Belemnia trotschi*—Arctiidae
99. *Mesotaenia barnesi*—Nymphalidae
100. *Pierella helvetia*—Nymphalidae

Alphabetical Species List

32. *Adelpha basiloides*—Nymphalidae
16. *Adhemarius ypsilon*—Sphingidae
83. *Aellopos ceculus*—Sphingidae
76. *Agrias amydon*—Nymphalidae
26. *Anaea aidea*—Nymphalidae
81. *Aphrissa statira*—Pieridae
79. *Archaeoprepona demophon*—Nymphalidae
43. *Arsenura drucei*—Saturniidae
9. *Astraptes* INGCUP—Hesperiidae
78. *Astraptes talus*—Hesperiidae
10. *Astraptes* YESENN—Hesperiidae
3. *Automeris belti*—Saturniidae
4. *Automeris io*—Saturniidae
85. *Automeris metzli*—Saturniidae
86. *Automeris phrynon*—Saturniidae
89. *Azeta rhodogaster*—Noctuidae
19. *Bardaxima perses*—Notodontidae
98. *Belemnia trotschi*—Arctiidae
61. *Bungalotis astylos*—Hesperiidae
62. *Bungalotis diophorus*—Hesperiidae
97. *Cacostatia sapphira*—Arctiidae
36. *Caligo atreus*—Nymphalidae
20. *Calledema plusia*—Notodontidae
87. *Cerura dandon*—Notodontidae
88. *Cerura rarata*—Notodontidae
11. *Chrysoplectrum pervivax*—Hesperiidae
42. *Consul fabius*—Nymphalidae
48. *Copaxa curvilinea*—Saturniidae
23. *Cosmosoma cingulatum*—Arctiidae
24. *Cosmosoma teuthras*—Arctiidae
60. *Creonpyge creon*—Hesperiidae
28. *Danaus plexippus*—Nymphalidae
72. *Dismorphia amphiona*—Pieridae
27. *Dysschema viuda*—Arctiidae
71. *Entheus matho*—Hesperiidae
84. *Enyo ocypete*—Sphingidae
22. *Epia muscosa* Apatelodidae
65. *Erbessa salvini*—Notodontidae
13. *Eueides procula*—Nymphalidae
45. *Ferenta castula*—Noctuidae
7. *Fountainea eurypyle*—Nymphalidae
74. *Gonodonta pyrgo*—Noctuidae

95. *Greta morgane*—Nymphalidae
50. *Hamadryas amphinome*—Nymphalidae
75. *Heraclides astyalus*—Papilionidae
25. *Historis odius*—Nymphalidae
53. *Hypercompe icasia*—Arctiidae
57. *Kloneus babayaga*—Sphingidae
52. *Letis mycerina*—Noctuidae
58. *Madoryx plutonius*—Sphingidae
41. *Mechanitis polymnia*—Nymphalidae
39. *Memphis mora*—Nymphalidae
40. *Memphis proserpina*—Nymphalidae
99. *Mesotaenia barnesi*—Nymphalidae
44. *Mimallo amilia*—Mimallonidae
18. *Mimoides euryleon*—Papilionidae
1. *Moresa valkeri*—Notodontidae
34. *Morpho amathonte*—Nymphalidae
33. *Morpho theseus*—Nymphalidae
96. *Mydromera notochloris*—Arctiidae
49. *Myscelia pattenia*—Nymphalidae
54. *Neonerita dorsipuncta*—Arctiidae
12. *Neoxeniades molion*—Hesperiidae
35. *Opsiphanes cassina*—Nymphalidae
5. *Oryba achemenides*—Sphingidae
6. *Oryba kadeni*—Sphingidae
38. *Othorene verana*—Saturniidae
59. *Oxynetra hopfferi*—Hesperiidae
47. *Oxytenis modestia*—Oxytenidae
92. *Parelbella macleannani*—Hesperiidae
17. *Parides iphidamas*—Papilionidae
37. *Pentina flammans*—Thyrididae
91. *Phocides nigrescens*—Hesperiidae
82. *Phoebis philea*—Pieridae
100. *Pierella helvetia*—Nymphalidae
46. *Porphyrogenes* BURNS01—Hesperiidae
80. *Prepona laertes*—Nymphalidae
68. *Proteides mercurius*—Hesperiidae
56. *Protographium marchandi*—Papilionidae
55. *Protographium philolaus*—Papilionidae
90. *Pseudodirphia menander*—Saturniidae
64. *Rejectaria atrax*—Noctuidae
93. *Rhescyntis hippodamia*—Saturniidae
94. *Rothschildia erycina*—Saturniidae
63. *Sarsina purpurascens*—Lymantriidae
8. *Siderone marthesia*—Nymphalidae
31. *Siproeta epaphus*—Nymphalidae
73. *Strophocerus thermesia*—Notodontidae
67. *Telemiades chrysorrhoea*—Hesperiidae
14. *Tithorea tarricina*—Nymphalidae
2. *Trosia* JANZEN01—Megalopygidae
77. *Urbanus belli*—Hesperiidae
30. *Xylophanes crotonis*—Sphingidae
29. *Xylophanes porcus*—Sphingidae
15. *Xylophanes tyndarus*—Sphingidae
51. *Zale peruncta*—Noctuidae
69. *Zaretis ellops*—Nymphalidae
70. *Zaretis itys*—Nymphalidae
21. *Zerene cesonia*—Pieridae
66. *Zunacetha annulata*—Notodontidae